THE NEW ANALOG

THE NEW ANALOG

Listening and Reconnecting in a Digital World

Damon Krukowski

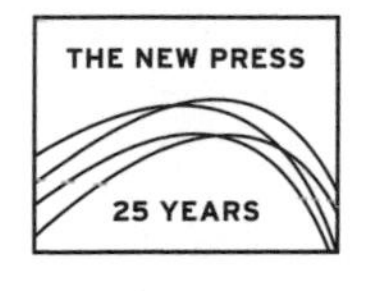

NEW YORK
LONDON

This book is made possible with support from Furthermore:
a program of the J.M. Kaplan Foundation.

Published in the United States by The New Press, New York, 2017
Distributed by Perseus Distribution

LIBRARY OF CONGRESS CATALOGING-IN-PUBLICATION DATA
Names: Krukowski, Damon.
Title: The new analog : listening and reconnecting in a digital world / Damon Krukowski.
Description: New York : The New Press, 2017. | Includes bibliographical references.
Identifiers: LCCN 2016045846 | ISBN 9781620971970 (hardcover : alk. paper)
Subjects: LCSH: Music--Psychological aspects. | Musical perception. | Sound--Recording and reproducing--Psychological aspects.
Classification: LCC ML3830 .K756 2017 | DDC 781.1/1--dc23 LC record available at https://lccn.loc.gov/2016045846

Book design and composition by Lovedog Studio
This book was set in Sabon MT

Printed in the United States of America

10 9 8 7 6 5 4 3 2 1

A project supported by the
Creative Capital / Andy Warhol Foundation
Arts Writers Grant Program

The inn we are directed to seems to us at first sight to be the dirtiest we have ever seen, but after a while it is not at all so exaggeratedly bad. It is dirt which is just there, that's all, and about which no more is said; dirt which will never change any more, which has made itself at home, which in a certain sense makes life more tangible, more earthly; dirt out of which our host hurries forth, proud towards himself, humble towards us . . . Who, one must ask, could still have anything on his mind against this dirt?

—Kafka, "The Aeroplanes at Brescia"

CONTENTS

THE NEW ANALOG

1

USER'S MANUAL

THANK YOU FOR READING this analog book. It requires no additional hardware, uses no power, and is 100 percent recyclable.

You will find that it is possible to read, or not read, any of this book's pages in any sequence. While its pages have been numbered sequentially to assist in navigation, there is no reason to consult these numbers if you do not wish. Should you like to highlight a passage, you will find that you can mark the page with most any implement at hand—even a fingernail will do. The paper of this book is also soft enough to be folded, torn, even shredded if that gives you satisfaction, without special tools.

You are free to share this book, resell it, or donate it to charity.

The author and publisher of this book do not have any information about you; they do not even know that you have a copy of this book unless they sent it to you personally. And if they did—send it to you personally, that is—you can always pretend to have read the book without having done so. You can also deny having read it, should that prove expedient. It's your business, really.

Welcome to the world of analog books!

Moore Meets Murphy

The oft-cited Moore's Law refers to the rapid development of integrated circuits since the 1960s—and therefore to computers and digital equipment generally—which follow a pattern of doubling in power and capacity every eighteen months.

But there's an overlooked corollary to this, which we might call Murphy's Moore's Law: if aspects of a given technology functioned better before the introduction of integrated circuits, they must be getting worse at the same fantastic rate. Twice as bad, every eighteen months . . .

Consider the typography of this book. In 1965, when Gordon Moore first formulated his observations about the rapid development of solid-state electronics, books were set in hot-metal type; that is, their words were cast into lead, resulting in crisp, detailed impressions on paper. What's more, the technology for hot-metal typography had at that point been refined by so many generations of designers and typesetters that even an inexpensive, commercially produced book like this would bear many marks of typographic excellence accumulated over time.

A few years later—while Moore was extending his law of

growth to personal wealth by cofounding the semiconductor manufacturer Intel—electronics began to make phototypesetting more cost-efficient than hot metal. Phototypesetting (or "cold type") was by comparison prone to distortion and breaks in letter forms, and limited in its ability to use the full range of delicate typefaces that had been designed over centuries for lead. But since it utilized electronics, the cost of cold type went down while its capacity rapidly increased, just as Moore observed. Over the centuries, hot-metal innovations had accrued at a speed somewhat closer to the flow of molten lead.

This is where Murphy comes in. Since cold type was in many respects lower in quality than what preceded it, increasing its availability could only lead to more and more bad typography. Which is exactly what happened. Today, any of us with a computer has the means to typeset, thanks to Moore. But only some are skilled at it, and as a result we are surrounded by a massive amount of typography without a minimum of professional standards. (Living with a graphic designer has made me acutely aware of this; public signage that fails to use smart quotes is among her bêtes noires.) Meanwhile, not only the commercial hot-metal typehouses but also their phototypestting successors have closed out of neglect—machines junked, the chain of skilled human expertise broken. The refined technology of hot-metal typography is limited now to "artisanal," specialty uses—a letterpress invitation to the retirement party for an Intel executive, say, but never an ordinary book like this.

Murphy's Moore's Law can apply equally well to a fast-moving, twentieth-century electronic medium like sound recording, as to a much older and more stable art like typography. In 1965, when even user's manuals were still set in hot-metal type, producer George Martin and his audio engineers at EMI Studios on Abbey Road were eagerly embracing any and every new electronic device for recording the most popular band in the world, the Beatles; as was their trans-Atlantic rival in recording mastery, Brian Wilson of the Beach Boys. *Revolver* and *Pet Sounds*, the innovative albums these two groups would release in 1966—eyes firmly on one another in competition and mutual admiration—remain widely acknowledged paragons of the art of studio recording. (The following year, the Beatles began to break apart while they struggled to top themselves with *Sgt. Pepper's Lonely Hearts Club Band*; and Brian Wilson did indeed crack up over his unrealized follow-up, *Smile*.)

If Moore's Law alone applied to sound recording, we would have exponentially better recorded albums today than were released in 1966. We don't. Rock and roll devotees who buy *Smile* bootlegs aren't the only ones who feel this way; classical music audiophiles treasure LPs from the same era, because they too have never been surpassed for quality. (A sprinkling of code words like "RCA Shaded Dog" and "Columbia Six Eye" are all one needs to open the door on that particular subculture.) As one typically prideful website devoted to classical music and high-end audio puts it: "'Collectors' have NEVER acknowledged any technological advancements in the production of records after around 1965."[1]

Did recording technology really peak in 1965, just as Gordon Moore was staring into his integrated circuits like a crystal ball? A lot of time, money, and hokum has been exhausted trying to preserve or reproduce studio conditions circa 1965: the microphones, the mixing boards, the tape decks, the amps, the tubes, the instruments—all command high prices and exude mythic mojo for musicians and audio engineers. Some of these are indisputably beautiful sounding, the likes of which have not been manufactured since. And thanks to certain enthusiasts, just as hot-metal type lives on in specialized use, it is still possible to record audio with these machines—just not for typical commercial purposes.

Obviously there have been countless innovations in sound recording since *Revolver* and *Pet Sounds*—both of which were intended to be heard in mono not stereo, for example. But since the introduction of integrated circuits, commercial recording and reproduction has, like commercial printing, followed Murphy's Moore's Law: it has doubled and redoubled time and again its capacity and speed, lowered and relowered its cost and availability . . . and decreased in quality. You needn't be an audiophile snob to conclude that today's MP3 downloads, or their streaming counterparts, sound worse than 1965's LPs—MP3s are *designed* to sound worse. It's a crucial part of what enables them to be so portable, cheap (if not free), and ubiquitous. That is, subject to Moore's Law.

What's So New About Digital?

One solution to Murphy's Moore's Law is to turn a blind eye to newer technology and do everything one can to keep the old ways going. It can be heroic—if quixotic—to stick to an outdated technology. The craft, ingenuity, and patience required to maintain older technologies is formidable, like those mechanics in Cuba who keep 1950s American cars on the road despite an embargo on U.S.-made parts *since 1962*.

Indeed, a defining characteristic of artisanal production is maintaining a technology in relative isolation. To operate a hot-metal letterpress today, or an all-analog recording studio, is to place oneself on a technological island with a dwindling number of compatriots who share the need for knowledge, parts, and skill to keep these machines going. Like a Cuban auto mechanic, you have been cut off from supplies by the surrounding industrial power as it moved on to newer paradigms that would leave yours, in the famous words of Trotsky, "in the dustbin of history."

An alternate solution is exceedingly familiar today from the strategy of "disruption" followed by so many digital-era enterprises, who urge us to break cleanly with the past and embrace the latest platform over earlier incarnations. Any hesitation at adoption only prolongs problems technology would have already solved, if we'd only get with the program.

This all-or-nothing response is extreme yet dominates popular discussion of the many anxieties provoked by the digital revolution. Op-ed pages and bestseller lists are filled with both condemnations and triumphant declarations of how technology is influencing every aspect of our lives, not least cultural production and consumption. Much of that discussion depends on the premise of a stark dichotomy: old v. new. In media, that would seem to equal analog v. digital.

But my experience as a musician doesn't jibe with that divide. Analog is not simply old, and digital is not only new.

Analog refers to a continuous stream of information, whereas digital is discontinuous. This distinction predates electronics, let alone integrated circuits. Any division of information into discrete steps is a digital process: from counting on our fingers, to calculating using an abacus, to (at least in some musicians' view) plotting notes on a staff of music.[2] Yet our senses remain resolutely analog. When we hear numbers counted aloud, see the beads of an abacus, or feel the vibration of a string, those sensations happen on a continuous scale.

Even unplugged, in other words, we find ourselves mediating between analog and digital. However ancient this process may be, the current paradigm shift from analog to digital for our communications is—as my career in music during this time can attest—very real, and moving very, very fast. When I first entered a recording studio, in 1987, there were no electronic digital tools in it, nor were any involved in the process of delivering music made there to its listeners. These short thirty years have been enough to witness that change completely.[3]

I began work on this book as an effort to understand better the terms of this change in the media I know best: sound and music. Digital life has no lack of keen critics, including many more scholarly accounts of its economic, social, and political ramifications. My focus is on our aural life and its cultural implications. Because for all the angst and boosterism surrounding the shift from analog to digital in the music industry, I feel the meaning of it has yet to be adequately described. It's as if we lack a vocabulary for articulating the changes I have experienced as both a producer and consumer of music—one of the reasons, perhaps, our conversations about it so often resort to unhelpful dichotomies of old v. new, or pro v. con.

To address this problem, the following chapters take up a series of processes that have changed for producers of recorded audio with the shift from analog to digital. Each of these is mirrored by a change in our relationship as consumers to the technology of sound. And each, I believe, has implications for our communications at large in the digital age.

"Headspace" looks at stereo hearing and raises the question of location—how we use sound to situate ourselves in analog and digital space.

"Proximity Effect" considers our use of microphones, and extends the question of location to the way we gauge social distance as we address one another.

"Surface Noise" focuses on sounds generated by audio media themselves, and discusses depth as an aspect of how we listen—what we hear when we listen closely.

"Loudness Wars" recounts the recent change in our use of volume, and more generally distortions of address and of hearing—the curves to our perception.

"Real Time" asks how sounds we exchange, under the different constraints of analog and digital time, make for a shared history or not.

Dumpster Diving in the Dustbin of History

Each of these examples is indicative of our changing relationship to noise.

Noise, to an electrical engineer, is whatever is not regarded as signal. Analog media always include noise, necessarily—efforts to minimize noise in analog environments adjust its ratio to signal ("signal-to-noise ratio"), but never eliminate it. Digital media, on the other hand, are capable of separating signal from noise absolutely. Given a definition of signal, the digital environment can filter out noise completely; this is fundamental to its efficiency as a medium for communications.

However, what I know well from working with sound and music is that *noise is as communicative as signal*. Which means Murphy's Moore's Law has been at work: consigning noise to the dustbin of history doesn't only multiply and speed our communications, it also diminishes them.

Nevertheless this book is hardly a Luddite's call. There are good reasons to dig into the dustbin of history, as Murphy's Moore's Law demonstrates. But especially when dumpster diving, one needs to be careful about what one pulls out. Uncritical nostalgia can lead to wholesale preservation of the bad as well as the good—an antique car, no matter how beautifully designed and constructed, is necessarily a gas guzzler. I see the digital disruption of our cultural life as an opportunity to rethink the analog/digital divide and re-examine what we've discarded—not in order to clean it up and put it back to use exactly as it was, but to understand what was thrown away that we still need.

A model I keep in mind for this kind of selective dumpster diving is Jane Jacobs, who in the early 1960s attacked the grandiose city planning of the postwar.[4] The push toward large-scale solutions to urban problems, based on Le Corbusier's Radiant City ideal and resulting in the myriad banal housing projects and shopping centers that aped it, was in Jacobs's view destroying the diverse, small-scale fabric of successful city living as it already existed. By resisting these new plans, she wasn't advocating the continuation of problems they were meant to solve, like street crime. Instead she saw the need to preserve ways in which those problems were already being addressed.

With the advent of the digital era, I believe that our cultural communications have entered a phase similar to the urban planning Jacobs railed against. Borrowing a page from *The Death and Life of Great American Cities*, *The New Analog* argues that we need to look more carefully at what was working well for us in the analog realm, so we don't destroy it in the rush of digital construction.

My poetry teacher Charles Simic once suggested that the old adage about translation be reconsidered: Poetry is not what gets lost in translation, as Robert Frost once said. Poetry may be precisely what *survives* translation. By the same token, I would argue that analog is not what is replaced by digital technology. On the contrary, analog may be precisely what *must survive* that transition.

Post-digital Analog

I first started writing publicly about these issues by sharing practical information with other musicians about the new paradigms for the music industry. In an article for the website Pitchfork, I discussed the specifics of what my first band, Galaxie 500, was now being paid by streaming services Pandora and Spotify and compared that to our traditional royalties.[5] "It would take songwriting royalties for roughly 312,000 plays on Pandora to earn us the profit of one—*one*—LP sale," I wrote. "On Spotify, one LP is equivalent to 47,680 plays." Using the example of Galaxie 500's first 7-inch single, "Tugboat," I calculated that "pressing 1,000 singles in 1988 gave us the earning potential of more than 13 million streams in 2012."

These numbers generated a fair amount of controversy, with some industry representatives insisting Galaxie 500's streaming royalties were either miscalculated or atypical, while other artists—some with far more commercial careers—began to come forward with nearly identical figures.

In the game of telephone that ensued, my article has at times been cited as a rallying cry against the new digital paradigm for audio, as if it could be boiled down to the single word "No." But in the article I had taken a different stance. For one, I made it clear that I myself am an eager consumer of streaming services because, like so many others, I enjoy the access it gives me to music of all kinds. (Although I did take the opportunity to point out that it would take 680,462 plays of "Tugboat" to earn back the cost of my annual Spotify subscription.) More importantly, the article was intended to make a larger point about what we are losing while we gain the convenience of streaming: the ability for the business of music to function successfully on anything but a massive scale.

That kind of problem reaches beyond the concerns of a given band and their royalty woes. To echo Jane Jacobs, I believe we are witnessing the death of great American media. The existing structures for music, publishing, film, radio, and journalism are being torn down as fast and as thoroughly as Boston's West End in 1958 and New York's Penn Station in 1963. Some of that destruction is inevitable, and some is likely for the good. But we need to understand more precisely what it is we are losing—like the ability of a young band to pool their money and press a single that will turn a profit even if it never reaches beyond their local circle of fans.

I don't intend to deny or fight Moore's Law, in other words. But I would like us to try and get the Murphy bit out of it.

This book is an effort to define aspects of the analog that persist—that must persist—that *we need* persist—in the digital era.

2

HEADSPACE

I SAW A WOMAN FALL from her bicycle in the middle of the street. "What happened?" I asked as I helped her up—the one car nearby had hardly come close. She took her headphones off and said, "I was totally self-absorbed. Suddenly I realized there was a car in the road. I braked and fell." The driver was there now, too, window down; he looked bewildered. She assured me she was OK, and continued on.

I rarely wear headphones outside because I feel vulnerable when I'm not alert to my surroundings—old instincts from a childhood in 1970s New York City. "Always notice who is around you—but don't meet their eye!" instructed my father. "And *never* fall asleep on the subway," my mother added.

But safely back in my bedroom, a pair of headphones was a de rigueur accessory to teenage life. The headphone album of the 1970s was meant to take you elsewhere—not into the street, but certainly out of your bedroom, where you were tethered to the stereo by a coiled cord like an astronaut tied to an orbiting spaceship. Close your eyes, turn up the volume, and fly into headspace.

Cut the cable, however—the one that kept my headphones in the bedroom—and it seems you risk confusing the external world with the internal one.

The woman I saw fall from her bicycle was disoriented *aurally*. Her hearing—a sense we use for localization—had been disrupted rather than aided by the technology she was using. That technology has a history both old and new, analog and digital. The oldest—stereo hearing—is a part of our evolutionary ability to locate ourselves in the world. The newest—smartphones communicating with global satellites—ostensibly does the same with greater accuracy and flexibility.

Yet even though the digital technology this woman was carrying could map her precise location on the earth, she could not *hear where she was going*. Her headphones had not only left the bedroom, her bedroom had left the building. She was occupying inner space—a place satellites can't map.

Localization Through Hearing

How did we get to this technological moment where, tethered each to our portable devices, we are walking, driving, riding until one day we pitch over in the street for no reason other than being "totally self-absorbed"? How did our means of communication go from bridging the distances between us, to isolating us in our own headspace?

To trace that story, we might begin in late eighteenth-century Milan. The great Teatro alla Scala—La Scala, for short—was built to replace the old royal theater under the auspices of Empress Maria Theresa of Austria. One of the prerogatives of funding a new theater is choosing the best seat in the house. For the empress, that did not mean one with the closest view of the stage but a box dead center at the back of the room: the only point where audio emanating from left and right are in perfect balance. You could say the entire theater pivots around her head. (Even a shockingly expensive seat to one side of the orchestra receives imbalanced sound from the stage—I am speaking from personal aural and financial experience.) [*Figure 2.1*]

In the Anglo-European concert hall, musicians are placed at one end and a central listening position facing them is nearly always considered best. One reason for these highly constrained conditions is our natural ability to locate sounds in space. Our two ears accomplish this by processing minute differences in audio arriving at either side of our heads; volume, timing, and tone all contribute to the sensation. Lower frequencies are the most difficult to locate, because their sound waves are largest in comparison to the size of our bodies—the difference in those waves heard from one side of our heads to the other can become too small for us to perceive. At the other end of the spectrum, in the range of the human voice

FIGURE 2.1

La Scala pictured from the stage. The royal box is at the back, sharing a centered left-right balance with only the conductor and a handful of cheap standing-room or upper-tier seats.

or the instruments that mimic it, our sound localization is acute. Hence the sensitivity to position for a listener at La Scala.

Binaural clues also play a role in how we pay attention to sounds. We pick a particular voice out of a crowd—or a single instrument from an orchestra—by focusing on sounds coming from its location and minimizing those arriving from elsewhere. Curiously, this ability to pay attention through localization is one that develops relatively slowly and degrades rather early in our aging. Researchers report that our capability for "spatial hearing" doesn't fully mature until age eighteen, continues to improve all the way into our forties, and then starts to decline after age fifty—as anyone in a busy environment trying to get a point across to the young or old is apt to have noticed.[1]

Hearing aids are notoriously bad at cocktail parties and crowded restaurants because amplifying volume doesn't improve spatial hearing—it only makes the same clump of indistinguishable noises louder. Localization is not a function of hearing *sensitivity*; it's the result of our ability to detect *difference* in what we hear from each ear. That is, it is dependent on our ability to listen in stereo.

From Proust to the Summer of Love

The significance of stereo hearing to our senses made it a focus of technological interest from the very beginning of electrical sound transmission. In 1881—only a few years after Alexander Graham

Bell's 1876 patent for the telephone and Thomas Alva Edison's 1877 invention of the phonograph—a stereo phone system was demonstrated in Paris for listening to the opera at a distance from the theater. Multiple transmitters at the foot of the stage relayed distinct signals to two handsets, one of which was held to each ear for a stereo experience of the live performance in progress. For a time, this "Théâtrophone" existed in Paris and a few other European capitals as a commercial service. Marcel Proust was a subscriber.[2] [*Figure 2.2*]

It wasn't until the late 1920s that engineers began to experiment with stereo recording, however. In 1931, a remarkable engineer named Alan Blumlein at the newly formed EMI in London published a paper that established patents for stereo recording, stereo records, and stereo disk cutting, all of which are applicable today. (A standard way to arrange two microphones for stereo recording is still called a "Blumlein pair.") By 1934, Blumlein had demonstrated the functionality of this new technology with a stereo recording of the London Philharmonic Orchestra performing Mozart's Jupiter Symphony at EMI's new Abbey Road studios.

Blumlein's bosses were apparently unimpressed. Stereo recording was shelved, and he was put to work developing a different new technology instead: television. (His patents in that field were also groundbreaking; the BBC would go on to broadcast the first public high-definition television signal in 1936.) Blumlein invented stereo records when he was twenty-eight years old, and died in a

FIGURE 2.2
Théâtrophone poster by Jules Chéret (1896).

World War II plane crash at thirty-eight. But he would have had to live past retirement to see that invention widely adopted. The first commercial stereo records weren't released until 1958, and it would take another ten years—until after the Summer of Love—before they became standard. The world of 1968 would have surprised Blumlein in so many ways, except hi-fi.

The Sound of One Ear Listening

Since stereo is so important to our hearing—specifically to our abilities to locate sound and to focus on one sound out of many—it seems a puzzle that it took so long for it to take hold in audio reproduction. Might it be because it's not that important to making and sharing music?

Revolver and *Pet Sounds*, the great rock albums of 1966 often cited as a paragon of popular music recording, were produced in mono even though stereo was available at the time. The Beatles did also release a stereo version of *Revolver,* but by many accounts the members of the band and their meticulous producer George Martin were not always present while it was being mixed, leaving the Abbey Road engineers to deal with it on their own. Stereo must have seemed like a faddish product, the kind of gadget that only a wealthy sybarite (like Proust?) might own. "The Stereo Scene" was the name of *Playboy*'s audio column, aimed at a man the magazine described as "in the midst of the biggest buying spree of his life. Cars, cameras, and hi-fi cabinets. Clothes, cognac, and cigarettes."[3] [*Figure 2.3*]

As for *Pet Sounds,* a stereo mix wasn't even attempted until thirty years later, when engineer Mark Linett was commissioned by Capitol Records to produce one as part of a four-CD box set documenting *The Pet Sounds Sessions.* Brian Wilson, the songwriting and producing genius behind *Pet Sounds*, participated in the compilation of that 1997 box set but not the stereo mix. He couldn't: Brian Wilson is deaf in one ear. As the liner notes explain, because of his hearing loss (most likely the result of a childhood beating by the Wilson brothers' abusive father), Brian Wilson "can't totally comprehend stereo."[4]

How did a man with hearing in only one ear produce one of the technically best recordings of the analog era?

Those same liner notes to *The Pet Sounds Sessions* give Brian Wilson's own explanation. Regardless of his hearing loss, "He always wanted his records to be in mono so that he would be in control of the listening experience. With mono, the listener hears it exactly with the balance that the producer intended. With stereo, however, the listener can change the mix, just by the turn of a balance knob or speaker placement."

That power of adjustment granted to the listener places each of us in the producer's chair. More importantly (in practice few of us actually reach for a balance knob), stereo limits the perfect place for listening to a space big enough for only *one at a time.* It places each of us in Empress Maria Theresa's box.

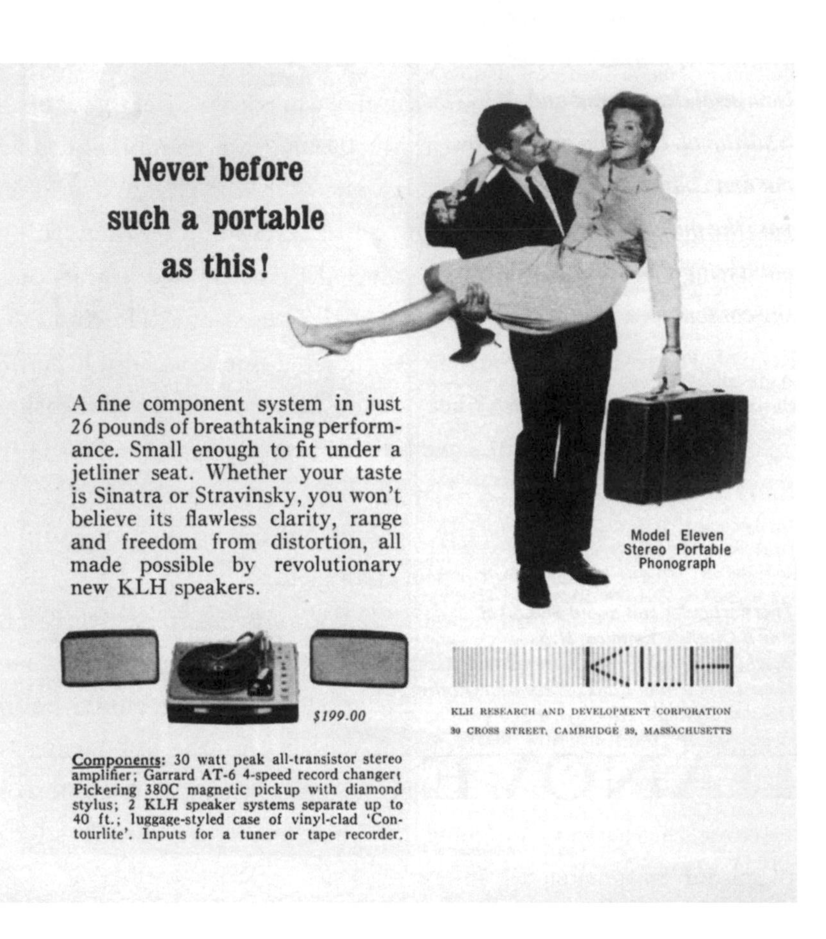

Never before
such a portable
as this!
A fine component system in just 26 pounds of breathtaking performance. Small enough to fit under a jetliner seat. Whether your taste is Sinatra or Stravinsky, you won't believe its flawless clarity, range and freedom from distortion, all made possible by revolutionary new KLH speakers.
Model Eleven
Stereo Portable
Phonograph
$199.00
KLH
KLH RESEARCH AND DEVELOPMENT CORPORATION
30 CROSS STREET, CAMBRIDGE 39, MASSACHUSETTS
Components: 30 watt peak all-transistor stereo amplifier; Garrard AT-6 4-speed record changer; Pickering 380C magnetic pickup with diamond stylus; 2 KLH speaker systems separate up to 40 ft.; luggage-styled case of vinyl-clad 'Contourlite'. Inputs for a tuner or tape recorder.

FIGURE 2.3

Stereo hi-fi ad from *Playboy* (October 1962).

Consider this seeming paradox, however: were the sound projected off the stage at La Scala a mono recording of an opera rather than the live event, the balance of the music would be the same throughout the house. This was what Brian Wilson didn't want to lose in the conversion to stereo. *Pet Sounds* is the same whether you are to the left of the speaker or to the right—whether you are on the beach blanket with it or on the next one over. It is made for sharing.

The air of exclusivity that ads for stereo in *Playboy* aimed to create is in fact one of its inherent qualities. It seems democratic to offer the empress's box to anyone with two speakers. But it comes with the proviso that each of us must occupy it alone.

Out of Our Heads

Let's return to my teenage bedroom, where, regardless of La Scala et al., stereo seemed very impressive indeed. By 1973, when Pink Floyd released *The Dark Side of the Moon,* the Beatles had broken up and Brian Wilson was living in seclusion (reputedly in the chauffeur's quarters of his own mansion). *The Dark Side of the Moon* sounded radically different to me than *Pet Sounds*, *Revolver*, or any of the other (mostly jazz) records I borrowed from my mother's collection, in part because it was so transporting on headphones. I was hardly the only one to make this discovery; teenagers everywhere slapped on cans and embraced this stereo album with such enthusiasm that it stayed on the *Billboard* charts for fifteen years.

One reason *The Dark Side of the Moon* became the ur-headphone album is that manipulation of space is as much a part of its effect on the listener as any of the elements to the music. It engages our sense of location—the same way we use stereo hearing in the audio world at large.

Indeed, the album opens not with music but with a heartbeat: a sound we wouldn't hear in a concert hall but rather inside our own heads. On the first track, "Speak to Me," voices and found sounds move freely between left and right, encircling us rather than projected forward from an imagined proscenium. The second track, "Breathe," behaves more like a regular recording with the band performing for us on a stage. But instead of ending with a cadence and pause for applause, it collapses and crossfades directly into "On the Run," where the sounds seem to get up from their fixed positions and wander around.

The effect of all this movement among sounds on *The Dark Side of the Moon* is cinematic—visual images are called up in the listener's mind, most obviously from programmatic elements like footsteps, snippets of dialogue, alarm clocks, and explosions but also from the continual reconfiguration of instruments around the listener's head. By the time a traditional song structure returns on track four, "Time," the sweeping synths, choral voices, and instruments themselves seem to be emanating from more than single points on a stage, leaving one floating in space rather than seated in the royal box. The title of the wordless song that completes Side A, "The Great Gig in the Sky," gives a name to this groundlessness. The band is still playing but the performance hall is far away, somewhere down on the diminishing earth.

No wonder some fans maintain that the album synchs with *The Wizard of Oz*.

If you listen to Side A of *The Dark Side of the Moon* in mono, as Brian Wilson would hear it, most of these spatial experiences disappear. What you are left with is a few classic rock songs embellished by rather inexplicable sound effects. (Pretty much the way I've always found Side B, which is disappointingly normal after all the fireworks of these first tracks.) *The Dark Side of the Moon* relies heavily on a manipulation of stereo that is heightened by the isolation and pure separation of left and right on headphones.

Pink Floyd and the album's producer, Alan Parsons, were so interested in the manipulation of space via multiple channels that they also produced a quadraphonic version—a kind of stereo squared. Widespread release of that effort was ultimately foiled by the problems quad faced as a commercial format,[5] but Parsons explained the strategy behind his mix for fellow audio engineers in a revealing 1975 article for *Studio Sound* magazine, "Four Sides of the Moon."[6] The basic stereo image for the album, he wrote, might be represented by this diagram:

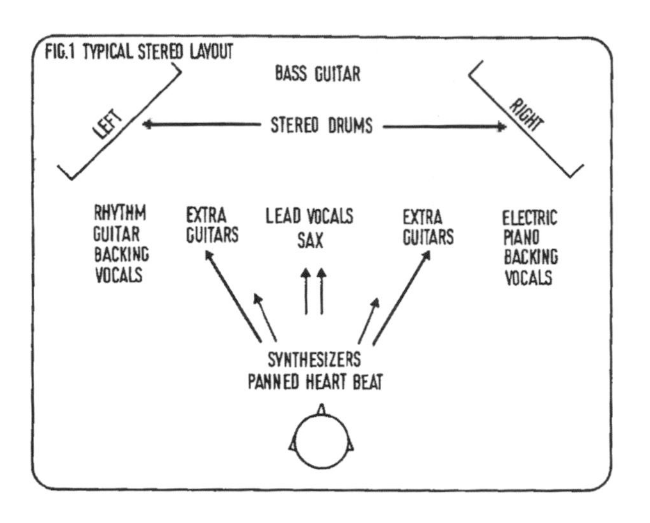

FIGURE 2.4
Illustration from *Studio Sound* (June 1975).

Apart from the trippy synthesizers and "panned heartbeat," this kind of arrangement is typical for stereo rock mixes and can be heard in the more conventional songs on *The Dark Side of the Moon*. Bass is placed in the center because low tones do not relate to our sense of location. Lead vocals are at the center because they are the focus of our attention; we face the singer head on to better listen to her or his words. Other instruments are panned to the right and left much the way they might be on a stage: drums in the middle but spread out so they envelop the band; piano off to one side, rhythm guitar to another; backing vocalists on either or both sides;

and solo instruments (saxophone, lead guitar) near or in the center, like the vocalist. The figure of the head with ears at the bottom of Alan Parsons's diagram could be—mutatis mutandis—Maria Theresa in her royal box.

But the quadraphonic version, Parsons explained, was meant to evoke a different kind of spatial map:

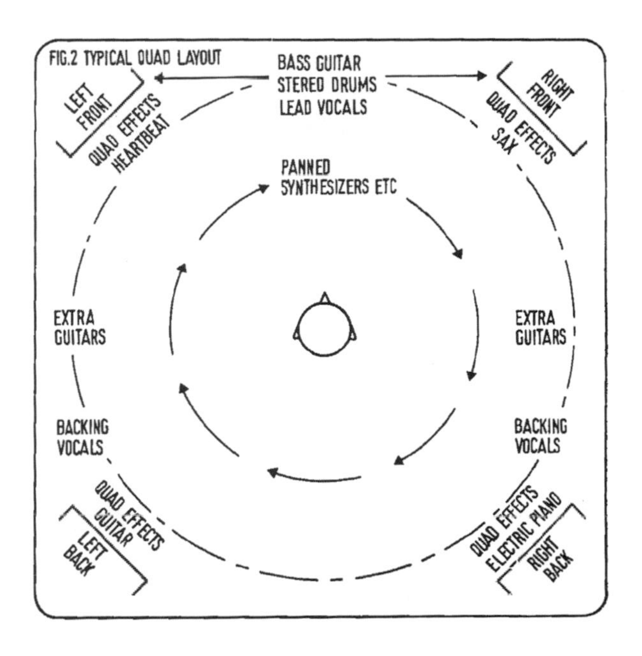

FIGURE 2.5

Illustration from *Studio Sound* (June 1975).

The head is still at the center, but instead of being at the back of the hall facing a stage, it is now surrounded by sounds that not only emanate from all quadrants but also are in motion.

Even though this quadraphonic mix is not the one most of us know, listening to the more experimental parts of the album on headphones communicates precisely this kind of enveloping space. That little head in the middle of Parsons's quadraphonic sketch is me and millions of other teenagers in their bedrooms circa 1975. Few of us were listening to Pink Floyd recordings as a simulacrum of hearing them perform onstage. Instead, we were each experiencing their music as part of an imagined space—a space not of the world but in our heads, conjured through the manipulation of our sense of location.

You Missed a Spot

For all its enveloping sense of a complete spatial world, *The Dark Side of the Moon* via headphones misses one key spot.

Stereo recording is—because of our naturally binaural hearing—more "realistic" in the sense that it better resembles our audio experience of the world. "Living Stereo" was the evocative phrase RCA used to describe their audiophile stereo LPs, because the sensation of listening with two distinct signals is like the listening we do in life, moving through space. Even the letters of the Living

Stereo logo seem to be so charged with vitality that they can't sit still, like antsy children.

FIGURE 2.6
Living Stereo logo for RCA Victor LPs (1958–1965).

But a stereo signal on headphones creates a blind spot, aurally speaking. The listener on headphones is forever at the center of the soundspace, just as Alan Parsons diagrammed for his quadraphonic mix of Pink Floyd and just as I experienced in my teenage bedroom. Sounds can enter that field from left or right, as we all know, and may even seem to move around us as they shift from one side to the other. But what we may not have noticed consciously is that the sounds we hear on headphones never seem to come from directly ahead.

The reason for that gap is our stereo hearing depends on the sum of information we receive from each ear, much the same as we see depth by summing together the perspective seen from each eye. When we listen to sounds in the world—and that includes sounds emanating from speakers—we hear every sound with both ears, using the differences between them to locate its source. But on headphones, the separation between left and right is absolute; we are using our ears *simultaneously*, yet *not together*. It's the aural equivalent of placing a block between your eyes. (Lining your hand up in front of your nose is enough to interrupt stereo vision, try it.)

As a paper written for the Institute of Electrical and Electronics Engineers (IEEE) describes it, "There are large errors in sound position perception associated with headphones, especially for the most important visual direction, out in front."[7] The resulting confusion—how could there be no sounds coming from directly ahead?—makes it "very difficult to externalize sounds," leading to an "inside-the-head sensation."

Physicist William M. Hartmann, who studies psychoacoustics, puts it this way: when wearing headphones, "The position of the image is located to the left or right as expected . . . but the image seems to be within the listener's head—it is not perceived to be in the real external world."[8]

What these scientists express about headphones and space is identical to what I learned listening to Pink Floyd. Headphones transport us, not outward but to internal space. Through an artificially pure separation of our left and right ears, they use our stereo hearing to remove us from the external world rather than position us in it.

How I Learned to Stop Listening and Love the Bomb

Consider the poor disoriented woman on the bicycle now. She is not only trying to map herself simultaneously in two planes of space—one via her bicycle, the other via her headphones—but she is doing so with a blind spot in front of her. "Distracted driving" is the least of it. Is it any wonder she wasn't sure exactly where she was?

Of course, there's an app for that. The same portable device interfering with the bicyclist's analog sense of location through hearing provides a digital replacement via its screen: the Global Positioning System (GPS).

GPS was developed by the U.S. Defense Department during the Cold War in order to better direct the trajectories of nuclear weapons aimed at the Soviet Union. Although the first satellite used for navigation by the U.S. military was launched as early as 1960, it took decades and untold billions of dollars before the current, completed system of information coordinated from twenty-four satellites went into effect. It took so long, in fact, that the Cold War had ended before it was fully operational—surely part of the reason that the United States decided to make this previously classified system available for nonmilitary use as soon as the last satellites finally came online in 1995. For the first five years, nonmilitary subscribers to the service were deliberately given less accurate information, but that practice ended in May 2000.[9] Now we all carry the full power of this massive military project in our pockets. [*Figures 2.7 & 2.8*]

So let's give our bicyclist some digital help mapping her trajectory. Instead of working with information gathered from her own analog senses, we might use data from these twenty-four GPS satellites orbiting the earth. Assuming everyone at that particular corner of the globe was equipped with smartphones (there's more than an excellent chance of that in the zip code where this occurred, 02138 in Cambridge, Massachusetts), it would be easy to digitally map the precise location of the bicyclist, the car she suddenly noticed in the road, and me on the sidewalk, observing.

In fact, none of us was on a collision course with any of the others. Had we each been looking at our digital devices for GPS-determined trajectories instead of calculating them using our analog senses, it's possible that the bicyclist would not have fallen, the car not have stopped, and this chapter never been written. If the only reason the three of us were not staring at our screens, despite each being equipped with billions of dollars' worth of military technology for positioning and trajectory calculation, is that our digital tools are still not sufficiently powerful for us to fully abandon our analog senses, then Moore's Law predicts they will be soon. (As I write, self-driving cars are being tested on the streets of Silicon Valley.)

But Murphy's Moore's Law points a different way. Perhaps you too have had the experience of following GPS navigation only to

FIGURES 2.7 AND 2.8

Frames from Stanley Kubrick's satire *Dr. Strangelove, or: How I Learned to Stop Worrying and Love the Bomb* (1964).

TOP: Headphones on, perhaps perusing the latest "Stereo Scene" column as well as the centerfold in *Playboy.*

BELOW: Slim Pickens guiding the bomb, analog style.

find yourself headed in a completely wrong direction or even arrived at the wrong address. In retrospect, the mistake is obvious: abandoning all analog clues to your whereabouts—signs, landmarks, the sun, maps, your internal sense of direction—can cause you to follow egregious digital errors, like locating the same number and street name in a different city. While traveling to that wrong address, your location was digitally mapped at every moment, more precisely than it ever could be using analog information. *And yet you never knew where you were heading.* The blue dot indicating your position was always at the center, while its destination remained offscreen. [*Figure 2.9*]

The position of that blue dot at the center of a GPS map is strikingly similar to Alan Parsons's diagram of the listener at the center of a quadraphonic Pink Floyd mix. In both situations—the headphone as the target of the mix, the blue dot as the target of the GPS map—the subject is fixed at the center point of a field in motion. The resultant perspective is literally, dizzyingly self-centered—or, to use the bicyclist's words, "self-absorbed." It's like a mechanized realization of Ptolemaic astronomy.

(Perhaps it's no coincidence that a technology developed to guide the destruction of life by nuclear weapons results in an egocentric map of the world?)

Just as our binaural hearing can be disorienting when it does not correspond to our actual movement in space, it is practically impossible to watch one's motion on the GPS map and simultaneously use vision to navigate through the world. Even were we able

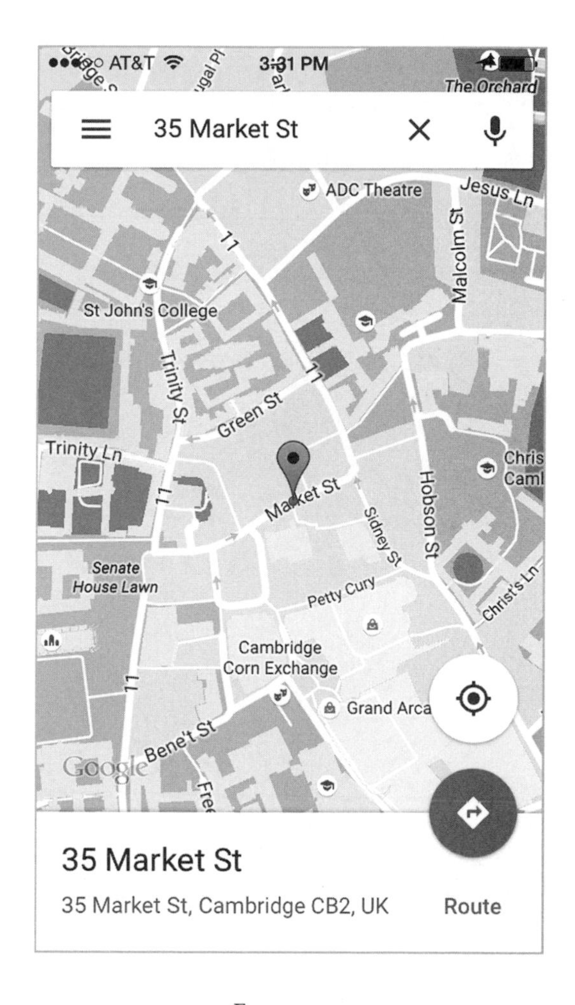

FIGURE 2.9

Google Maps returned this location for a friend's address at 35 Market Street, Cambridge, placing it in England rather than Massachusetts.

FIGURE 2.10

The Ptolemaic system as pictured in the *Margarita Philosophica* (Pearl of Wisdom) of Gregor Reisch (1503).

to watch both at once—as future technology will no doubt make possible—being in motion and simultaneously mapping ourselves through external data leads to disorientation.

Without analog clues to location—information gathered from our own senses—the world becomes an Alice in Wonderland–like place where signs pointing north nonetheless lead south, and sounds come from the left or right but never straight ahead.

And where we are each at the center of the universe. [*Figure 2.10*]

In a Lonely Place

Funny thing about the center of the universe: it's a lonely place.

There's even less space between a pair of headphones than at the perfect center of a stereo mix like the royal box at La Scala. If you are dancing while wearing earbuds, like in the iconic Apple "silhouette" ads for the iPod, you dance alone. The "inside the head" sensation that psychoacousticians assign to the aural blindspot in headphones has another, very simple explanation: it's the only place the music is audible.

GPS navigation is similarly isolating. Given a map that *follows you*, rather than the other way around, there's never a need to ask for directions. Who could possibly know more about where you are, and where you intend to go, than you? Another person could only give you their perspective—a view from their map, with themselves at the center.

In other words, once our maps are tailored to each of us with pinpoint accuracy, are we ever truly in the same place as one another? Can we still share a map?

Perhaps no more than we can share a pair of earbuds—which is possible, but only for a maximum of two. And even then, you won't each hear the same signal . . . unless you listen in mono.

I Can't Live Without My Radio

Time was, playing music necessarily meant sharing music. Sound travels.

In the first part of the twentieth century, recorded and broadcast music replaced parlor instruments in the home, and "playing music" began to mean listening together rather than actually playing together. Yet it remained a group activity—even if not everyone in the group wanted it to be. Early phonographs are *loud*, and there are no volume controls on Victrolas. (Hence the phrase, "Put a sock in it.")

Transistors made battery radios portable starting in the mid-1950s. But it wasn't until the end of the 1960s that FM stereo broadcasting and cassette tape brought loud recorded music into the street. The longer waves of the bass frequencies available on FM and via tape carried much farther than tinny transistors tuned to AM, and manufacturers of portables played up that novelty by adding more batteries and bigger, and bigger, and bigger speakers—until some of these "boomboxes" could only be carried by the largest among us.

My radio, believe me, I like it loud
I'm the man with a box that can rock the crowd
Walkin' down the street, to the hardcore beat
While my JVC vibrates the concrete[10]

By October 1972, New York City had passed a noise ordinance prohibiting "the playing of radios, phonographs or tape recorders on public transportation."[11] Hence James Todd Smith aka LL Cool J's boast on his 1985 single "I Can't Live Without My Radio" that his "JVC vibrates the concrete" and he plays it "everyday, even on the subway / I woulda got a summons but I ran away." [*Figure 2.11*]

Terrorizing my neighbors with the heavy bass
I keep the suckas in fear by the look on my face
My radio's bad from the Boulevard
I'm a hip-hop gangster and my name is Todd
Just stimulated by the beat, bust out the rhyme
Get fresh batteries if it won't rewind
Cos I play everyday, even on the subway
I woulda got a summons but I ran away
I'm the leader of the show, keepin' you on the go
But I know I can't live without my radio

If you weren't on a train with someone gangsta enough to risk a noise summons, the New York City subway was largely music-free in the 1970s. Headphones were legal—but, as my parents warned me, stopping up your ears to your surroundings was unwise. Anyway, who would use headphones with a boombox? The whole point, as LL Cool J explained, was to play it loud enough for everyone to hear: "Terrorizing my neighbors with the heavy bass . . ."

Filmmaker Chantal Akerman captured this "silent" era of the New York subway in the course of making her diary film *News from Home*. Several extended shots observe a single car from a fixed point of view during the summer of 1976. Looking at Akerman's movie now, it is striking how anchored the subway riders are in that particular space. A few read, but most just look around—at one another and at Akerman. There were plenty of reasons to distance oneself from the immediate experience of a hot subway car in the summer of '76, but there simply weren't the technological means of escape that we all carry with us now. You might say that everyone sweating there together knew exactly where they were, for better and worse. [*Figure 2.12*]

The Sony Walkman began to bring headphones into the street as early as 1980, but it took smartphones to stick them in all our ears. Digital communications now keep us constantly plugged in to a place other than the one we are in—whether via phone, internet, or recorded music. Screens and earbuds are the points of contact.

Picture Akerman's subway car today (you may be on one now). How many riders remain in just a single time and place? GPS might

FIGURE 2.11
LL Cool J with his radio (1985).

FIGURE 2.12

Frame from Chantal Akerman's *News From Home* (1976).

locate each of us in the same car. Yet the stream of digital information can put each of us in a different space than the others, even as we hurtle together through a tunnel on fixed tracks. Crowded in the subway, any of us might be as close or closer to somewhere or someone not on the train at all. That is, we occupy the space *simultaneously* but *not together*—the same discontinuous way our ears experience location on headphones. Our aural blind spot may well extend to one another.

Bring the Noise

In audio, a solution to the blind spot we experience on headphones is "crosstalk"—feeding a part of one channel's sound into the other in order to re-create the kinds of binaural clues we use to locate sounds in space. Crosstalk is in fact nearly unavoidable in the world at large, because the pure left/right separation we routinely experience through headphones can happen only in a highly rarified physical situation . . . like wearing headphones. Pipe it back in, and we feel more like we are listening to sounds in the natural world.[12]

Yet crosstalk is viewed traditionally by engineers not as a boon to perception but as something to be eliminated in search of clarity. Crosstalk is "noise" to be separated from whatever "signal" we are after. In information technology, the highest possible signal-to-noise ratio is considered ideal—which is why crosstalk is considered a problem in telephony, wireless communications, integrated circuits, and, yes, audio.

At the beginning of this chapter, the concept of spatial hearing was introduced in relation to stereo. Locating a sound in space is an aspect of how we pay attention to it—by focusing on that particular audio as signal and filtering out sound coming from elsewhere as noise.

But notice how spatial hearing depends on a constantly shifting definition of signal and of noise. If I want to listen to the person across the table from me in a restaurant, I block out the noise from the rest of the room. If I want to listen in on the conversation at the next table, I tune out the talk at my own. In other words, spatial hearing is dependent on the *presence of noise as well as signal.* If everything were signal, the restaurant would be a screaming mass of sound, and we wouldn't be able to focus our attention on anything at all. (This is exactly what happens during childhood, before our sense of spatial hearing is fully developed, and what can happen again as our spatial hearing deteriorates in old age.)

This constant calculation is something we see on the faces of Akerman's subway riders. They each look around the car, gathering what they need to know about their surroundings—filtering out the signal from the noise. There's nothing external to consult (this being New York in the 1970s, there aren't even recorded announcements on the train). There's just the information received by analog senses of eyes and ears that needs to be evaluated and sorted—exactly what our digital devices can now do for us more quickly, and in many ways more accurately.

But, as we've seen, not in every way.

Consider again the problem of misleading GPS instructions—locating Market Street, Cambridge, on another continent—and how easily that can be averted by simply opening one's eyes to the analog information all around. But like the noise of crosstalk, many analog clues to our location are contradictory and may even seem to reduce our ability to navigate. Signs are confusing, the road looks alternately familiar and strange, the map is vague about the one thing we need to know from it . . . Why else do we argue in cars so much?

Just as with spatial hearing, our sense of location as we move through space is dependent on a constantly shifting definition of signal and noise. Digital information can tell us where our blue dot is, precisely, at any given moment. But it's grappling with our uncertain position in the analog world that gives us a sense of where we are heading.

In the end, this is I believe what happened to the bicyclist I saw fall in the street. She surrendered her shifting analog sense of what is signal and what is noise, replacing it with the digital stream of information on her headphones—a stream that is *signal only.*

It wasn't that she had stopped paying attention. The problem was that she had stopped paying attention to noise.

3

PROXIMITY EFFECT

I'M GOING TO LEAN IN CLOSE as I say this: *noise is as communicative as signal.*

Back when phones were heavy, black, and tied by a cord to a wall, like so many others I spent an inordinate amount of time on them. "On them" didn't necessarily mean speaking so much as listening; once the phone was off the hook, it was a continual receiver regardless of how much talking actually went on. As a teen, I hung out on the phone the way a previous generation had hung out on the corner. Talking is a key ingredient to hanging out, naturally, but so is not talking. Especially if those not meant to hear your conversation might be listening, like your parents in the same or next room.

Having an extension in your bedroom—a substantial extra expense in the days of wired telephony, hence a luxury—was a key into more private talk but also more private silence. Silences on a publicly positioned phone might be a necessity, but silence on a private phone could be an intimacy.

The ultimate privacy, and therefore intimacy, was a number of one's own. The "private line" is a typical plot device from movies of the time: the politician's phone for those with special access; the businessman's call that doesn't go through the secretary; Joan Crawford's line direct to the tub in *The Women*, which Rosalind Russell discovers is "not an extension" and therefore expressly for a lover. [*Figure 3.1*]

It's not always what is said on the private line that moves the plot in these films—a private line was proof of intimacy even when nothing is being said. Especially if it's off the hook and anything at either end can be heard. The key scene in Hitchcock's *Dial M for Murder* is a phone call placed not for talking at all, but listening only. [*Figure 3.2*]

What Ray Milland hears at the other end of the phone is *noise*. The same was true for less dramatic calls placed on those heavy old analog phones, including my teenage ones. The rustling of clothes. A pencil doodling on paper. Breathing. The presence of another.

Mobile phones make dreadful listening devices for noise, by contrast. "Are you there?" is a constant refrain in cellular conversations, because it's so difficult to sense another's presence when they're not busy speaking. The silence on a cell phone is what audio engineers call "digital black." Digital black is not just the absence of signal but the *absence of noise*. It's nothing like the evocative

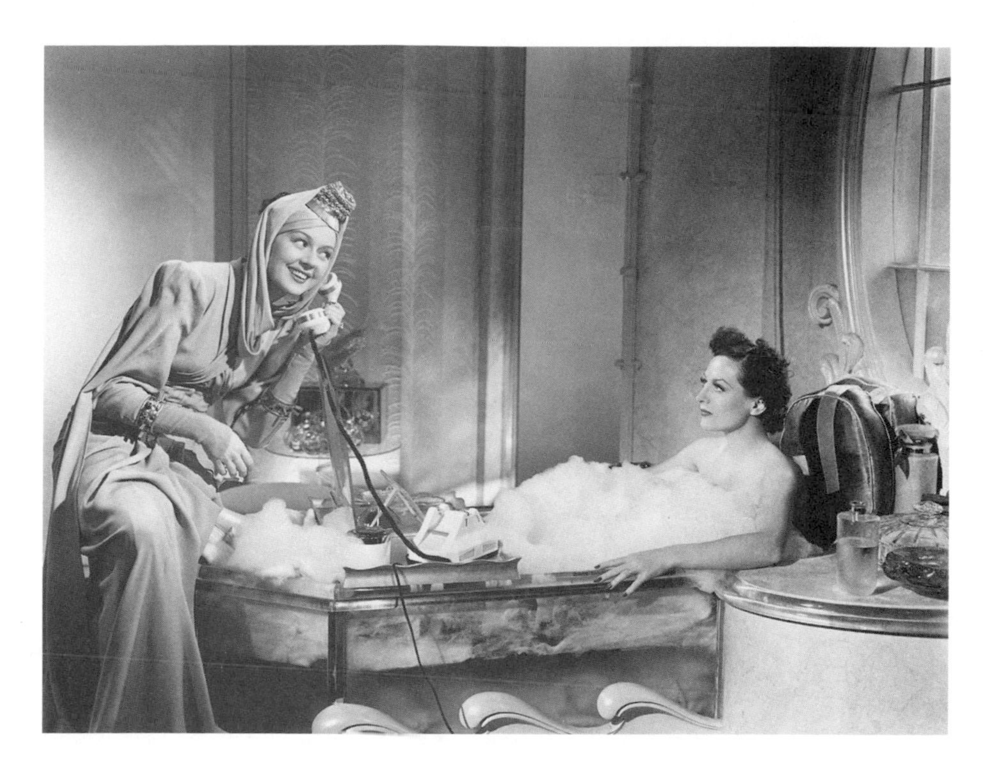

FIGURE 3.1

Rosalind Russell on the phone and Joan Crawford in the tub, production still for *The Women* (1939).

FIGURE 3.2

Ray Milland on a pay phone in a frame from *Dial M for Murder* (1954).

silence of Joan Crawford smoking in her bath or Grace Kelly being choked by a hit man.

Can You Hear Me Now

Phones were in fact invented as listening devices as much as speaking ones. Alexander Graham Bell was a teacher of the deaf—his mother was deaf, and he eventually married one of his deaf students—and his experiments leading to the patent of the telephone were directed primarily toward the ear rather than the mouth. Bell's original telephone successfully transmitted sounds across distance, but speech had to be shouted to be intelligible. Bell invented, in effect, an electric hearing trumpet. [*Figure 3.3*]

However, shouting doesn't bridge the distance between people so much as emphasize it. Bell's invention had little potential for intimacy until it was combined with a different inventor's brainchild: Thomas Alva Edison's carbon microphone. (Or perhaps it was Emile Berliner's carbon microphone; the patent dispute dates to 1877 and, like all such disputes, will likely never be resolved.) Edison, it is interesting to note, was himself hard of hearing. "Broadway is as quiet to me as a country village is to a person with normal hearing," he told his biographers. "In experimenting on the telephone, I had to improve the transmitter so I could hear it."[1] [*Figure 3.4*]

A microphone amplifies not only what we say through it—the signal—but everything around that signal, which sound engineers call *noise*. And the engineers for digital signals have developed a suite of tools to eliminate it.

Edison had no such tools available for his carbon microphone. It is an exceedingly simple device, yet proved so effective for telephony that it remained in place from the time of its first adoption in the 1890s all the way through my teenage conversations in the 1970s. It wasn't until the 1980s (and the breakup of "Ma Bell") that electret microphones—the same miniature mics we now use for cell phones—finally began to supplant Edison's original design. [*Figure 3.5 & Figure 3.6*]

Despite their tiny size, electret mics are more sensitive than carbon mics, not less—I use one mounted inside my acoustic guitar for amplification on stage because it captures such a full range of tones from the instrument. Carbon mics, by contrast, have a very limited frequency response. Think of the stereotypical telephone sound applied as an effect to voices in certain songs, like "Uncle Albert" from Paul McCartney's album *Ram*. (On that track, McCartney also vocally imitates the ring of an old English phone by burbling; his "ringing" starts 1:10 into the song, and the voice on the other end finally picks up at 1:30.) The phone effect is made in the recording studio by filtering a vocal track to eliminate both its upper and lower frequencies, leaving a bluntly truncated, tinny-sounding slice in the middle—the only part Edison's carbon mic transmits.

FIGURE 3.3

Alexander Graham Bell at upper right, posing in 1871 with fellow teachers and students at the Pemberton Avenue School for the Deaf in Boston—an institution still in operation as the Horace Mann School for the Deaf, the oldest public school of its kind.

FIGURE 3.4

Thomas Alva Edison cupping his hand to his right ear, in which he had partial hearing; he was completely deaf in the left.

The manufacturing breakthrough for electret mics took place in 1962, when engineers at Bell Labs discovered that metalized Teflon foil could provide a permanent static charge, enough to power a microphone through nothing but the materials of its own construction. This eliminated the need for an external power source and allowed high-quality sound reproduction on a mic small enough to be mounted inside a guitar, clipped to a lapel, or built into the handset of a phone that could now shrink around the mic as much as its other elements allowed. The electret is such an effective and useful design that more of them are currently manufactured—some estimate billions each year—than any other type of microphone.

Why, then, do we seem to hear less on the phone now than when they used carbon mics?

Plain Old Telephone Service

In the digital era, the sound of our voices on the phone is not determined primarily by the microphone, but by the way that sound is treated before it reaches our ears. It is the phone *as a listening device* that has altered so radically.

"Plain old telephone service" is how the analog era of telephone transmission is often now referred to—POTS for short. POTS carried our voices on copper wires through various generations of exchanges (switchboard operators being the most cinematic, though the direct dial tone may have left the most lasting aural impression), along collective "trunk" lines, back through "branch" exchanges until they reached their destination at our interlocutor's ear. The sound input via the carbon mic might travel a great distance, but if all went well along the line it arrived more or less the same as it started out.

With the advent of modems and fax machines in the 1980s, POTS had to make room on the lines for data as well as sounds entering the system via microphone. At first, modems and fax machines used acoustic couplers to translate data to sound so it could be carried along the old copper wires the same as a voice. Only so much data, and only so quickly, could be transmitted in this analog manner, however. The more efficient solution engineers hit upon was to stop converting data to sound, and instead convert sound to data. This spelled the end of POTS. Integrated Services for Digital Network (ISDN) was codified in 1988 but was soon itself supplanted by broadband, Voice over Internet Protocol (VoIP), and—most pervasively for voice communication—the digital cellular network we have now come to rely on for so much of our phone service.

The first step in making all that possible is transforming our voices into data; that is, they must be digitized. A microphone—*any* microphone, from Edison's carbon button to the cell phone's miniature electret—is an analog device. All microphones work on the same principle: sound pressure moves a diaphragm, whose fluctuations are converted to an electrical signal. That electrical signal is quite literally an analogue for the sound wave that created it, and

FIGURE 3.5
A photo of one of the earliest carbon button mics, filed for patent with the Library of Congress in 1877.

FIGURE 3.6

Carbon mic removed from the handset of a Western Electric model 500 telephone, designed in 1949 and manufactured until the end of the plain old telephone service (POTS).

thus easily reconverted to sound by turning the process around—a loudspeaker is the inverse of a microphone, running electrical current through a diaphragm to create sound pressure. (Edison's carbon mic could actually serve as a loudspeaker too, so the first exchanges simply placed them face-to-face in order to amplify the signal down the line.)

Cut that direct electrical signal between mic and speaker, and you cut the analog transmission of sound. Hence all those telephone poles, stringing our mics and speakers together. POTS was a hugely extensive but very direct system of sound transmission, not unlike an analog recording studio's patchbay connecting inputs and outputs via cables.

Digitizing sound takes the electrical signal generated by an analog microphone and converts it into zeros and ones; in the recording studio, this process is referred to as A-to-D. Once sound is D, it can be treated the same as any other digital information—no telephone poles, no patchbay, no direct connection of any kind required. That's because the transmission of digital data is *discontinuous*, in time as well as space—its reconstruction at the receiving end doesn't depend on a continuous analog stream of the original signal, like POTS. The last bit of a given set of digital information can arrive first, if need be, and each bit can take any available path to its destination regardless of the others.

The internet, remember, was designed for military communication in the wake of nuclear attack. POTS could be taken out by a pair of wire cutters. Which makes it all the more remarkable that it lasted through a century of use.

Cutting Ties

The stability of POTS from Bell and Edison's time through the 1980s was remarkable but perhaps not mysterious. Its system of "person-to-person" calls was, for all the twentieth-century electric wizardry of sending a voice down the lines, a very human and direct connection across space. Even children play telephone by whispering into one another's ears, moving a message around a room by converting sound pressure from voice to ear and back again. (The English call this Chinese whispers—because the message becomes as unintelligible as Chinese, or because, given enough children, it might travel all the way to China?) POTS was an electric amplification of our acoustic means of person-to-person communication, extending the reach of our whispers into one another's ears.

Just as in the children's game, the linguistic message sent down analog phone lines wasn't always easily understood. Comedians made use of this slippage right away—"Cohen on the Telephone" was a popular routine released multiple times on records from 1913 on, using Yiddish-inflected English to punch up the joke of being misheard over the line. ("I am your tenant, Cohen . . . No, I'm not goin', I'm stopping here awhile . . .")

In the movies, the phone soon established itself not only as a useful plot device but as a locus for comic misunderstanding. A telephone exchange between the great character actors Eric Blore and Edward Everett Horton in the Astaire-Rogers film *Shall We Dance* (1937) develops into wordplay worthy of the avant-garde literary group OULIPO:

> BLORE: I'm at the Susquehanna Street Jail. Susquehanna. Sus-que-hanna. S-U-S-Q . . . Q! Q. You know, the thing you play billiards with. Billiards! B-I-L-L . . . No, L, for larynx. L-A-R-Y-N . . . No, not M, N! N as in neighbor. Neighbor, N-E-I-G-H-B . . . B. B! Bzzz. Bzzz. You know, the stinging insect. Insect! I-N-S . . . S! S for symbol., S-Y . . . Y! Oh, Y . . .
>
> HORTON: Well, why? Don't ask me why!

Still, for all it lacked in sound quality, POTS successfully communicated *distance* loud and clear. When you picked up an analog phone you completed a physical circuit from the voice at the other end to your ear, with all the space between measured by copper wire. Noise on the line was understood as an artifact of that distance; any misunderstandings it caused were only further evidence of it.

Eliminate that length of wire, and perhaps you eliminate our sense of distance as well. Despite our cell phones knowing our precise location via GPS, whomever we call on them has no idea where we are: "Where are you?" has become as common a way to answer the phone as "Hello." Even 911 emergency operators have to ask where a call is coming from. (For those who grew up talking through copper wires strung overhead, "What town are you in?" is a very disconcerting reply to a local call for help.[2])

Of course, there's a similar vagary of geographical location on many nonemergency calls these days, when "local" customer service numbers pick up continents away. In my own experience, this spatial disconnect has at times reprised the hoary old routine of "Cohen on the Telephone," though rarely for laughs.

Mobile Phone Booth

On a cell phone, there is no indication whether a given call is arriving from another country or from the next room. They look the same (now that our numbers travel with us, area codes are useless for determining current location), and more importantly they *sound* the same.

To make an analog call intelligible required relative quiet—the public telephone booth (or box, in British parlance) was equipped with a door not just to help Clark Kent surreptitiously change into Superman but to shield the phone from the noise outside. [*Figure 3.7*]

The carbon mic may hear only a limited range of frequencies, but it transmits everything within them—signal and noise. True

FIGURE 3.7

Clark Kent's first use of a phone booth for this purpose, from the Sunday newspaper strip, December 27, 1942.

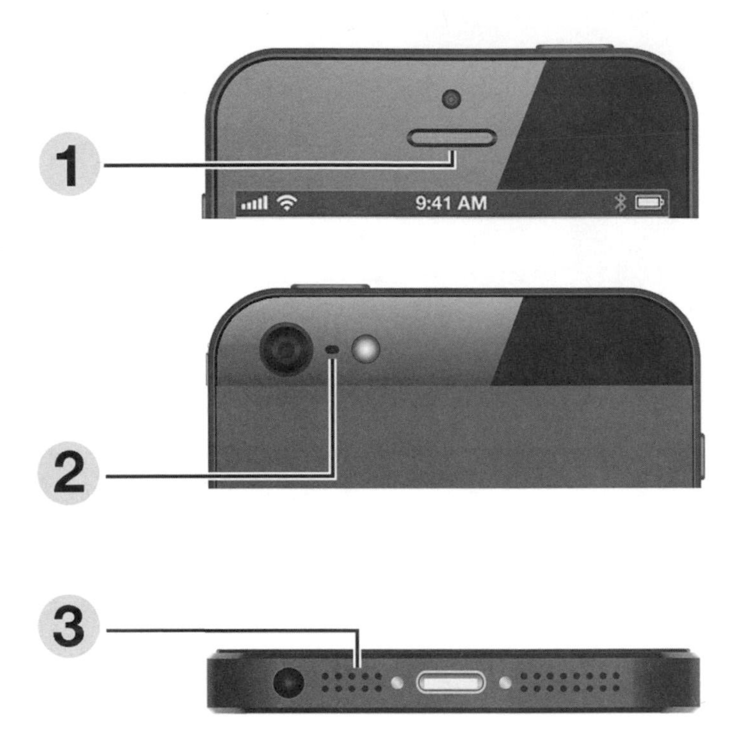

FIGURE 3.8
The three mics of the iPhone.

to Bell's original interest in the phone as a listening device, the analog phone simply opens a mic in a given space. The speaking voice in that space might be the loudest sound we hear, given its placement closest to the mic, yet it remains one sound among many. As a teen, although I could just manage to get the kitchen phone out the doorway and into the next room by pulling the cord its full length, the sounds of my family around the table were still audible at the other end. No Superman-like changing of identities was possible.

The digital phone, on the other hand, isolates our voices in a manner even the most private phone box never could. It accomplishes this despite the fact that the electret microphone is more sensitive than Edison's carbon mic, capturing more of the surrounding sound to a given signal. In fact, the current iPhone actually has three of these microphones: in addition to one at the bottom, by your mouth, there's one at the top (by your ear), and one around back (by the camera lens). [*Figure 3.8*]

As any audio engineer can tell you, two mics are all you need for stereo recording, and one does the job just fine in mono. So why three on the iPhone, which transmits our voices in mono regardless? These additional mics are there not to capture existing audio—the analog sounds we hear in the world—but to help process those sounds as digital data. They are used to isolate signal from noise.

Noise Cancellation

Among the technologies that the additional mics on the iPhone make possible are "noise cancellation"—recording sounds in order to negate them—and "beamforming," a system for preferential listening from a particular direction.

Noise cancellation works by generating an equal but opposite wave to a given sound in order to negate it—the two waves then sum to zero. In the late 1980s, the audio company Bose first incorporated this engineering into a headphone set for airline pilots, and it wasn't long (Moore's Law) before the technology became cheap enough to market to consumers. Noise-cancelling headsets built by Bose or any other company use a built-in microphone to listen to the ambient sound around the user. The polarity of that sound is then reversed and actively generated by electronics, leaving any sound not measured by that mic untouched. Thus it sorts "signal" (sound arriving from inside the headphones) from "noise" (sound arriving from outside the headphones).

Beamforming is a complementary technology that employs multiple microphones to locate the source of a signal, much the same way we use our ability to listen in stereo for spatial hearing. If the source of a given sound—a voice directed at a speakerphone, for example—is arriving from a particular direction, beamforming can focus on sounds from that location as signal and treat those from other directions as noise.

Both these technologies work to define and isolate signal. On a phone, that signal is the voice—as opposed to the voice's family around the table in the next room. A digital phone is capable of isolating our voices even more effectively than a phone booth, because the line it draws between signal and noise is not merely straight like a door. Instead that line can be irregular, flexible, and adaptive, conforming to the ever-changing sounds around us as it nimbly erases them.

Perceptual Coding

It's not only the atmosphere surrounding a voice that is removed from the sounds on a cell phone. It's also the atmosphere of the voice itself.

Perceptual coding is the term engineers began to use in the late 1980s for applying psychoacoustic research to digital sound processing. What perceptual coding makes possible is eliminating not merely the noise framing a signal, but those parts of the signal itself that are unnecessary for communicating data. If the voice on a phone is intended to communicate words, why not narrow the definition of signal to *just the words* in order to improve the accuracy of their transmission? The rest of the voice—those aspects that do not help a listener understand the words—can then be separated out and reclassified as noise.

What perceptual coding achieves is a more precise definition of signal, and thus a more efficient transmission of data. It helps make it possible to reduce the large sound files on CDs to portable or streamable MP3s. And it makes it possible to communicate over a cell phone from the worst environments for intelligible speech: in a car, at a bar, on the subway. Think how difficult it can be to talk across the table in a crowded restaurant, yet how relatively easy it is to be understood from a cell phone in that same environment. (You might try calling your dinner partner next time.) The difference is perceptual coding—your voice on the cell phone has been converted into data that contains only what is necessary for its words to be understood.

Which leaves a lot out.

When Apple marketing VP Phil Schiller publicly introduced the new and improved audio features for the iPhone 5 in 2012, he explained in his pitch, "The data in your voice . . . doesn't sound entirely natural all the time."[3] That is, the analog sounds of our voices—their "natural" qualities that we hear when we speak (or sing) to one another in a given environment—are not the same as what makes them intelligible as digital data.

One way to experience that difference is to turn off the digital sound processing applied to the mics on an iPhone. There's no option for that in Apple's iOS, but audio engineers have programmed workarounds. The podcast app Bossjock Studio includes this option, for example, and its designers have helpfully posted sound files with the iPhone's processing turned on and off for comparison.[4]

With the iPhone's digital sound processing on, the voice is restricted to a narrow range of volume and frequencies. The result is intelligible for content—*what* is being said is very clear—but *how* the message is delivered is lost. Is the voice loud or soft? Are we being addressed intimately or publicly? Can we hear hints of other meanings in the speaker's voice, or does the delivery match the words exactly?[5]

Answering any of these questions is more within reach when listening to a voice without digital processing. Even with a banal test message, it's possible to form an impression of the speaker: intonation, pauses, and emphasis are individual quirks that register much more clearly without the perceptual coding of digital processing.

When we use a cell phone to communicate, we use the digital assistant Siri's ears—we listen for intelligibility. But when we use an analog microphone (and an iPhone mic without digital processing is, like all microphones, an analog device) we use *our* ears, which are adapted and accustomed to the full range of the voice: the parts that make it intelligible and the parts that don't. Among the parts that don't are those that make a voice sound "natural," as Schiller put it.

An audio engineer might articulate that differently, since there's no natural in the studio though there are many ways to focus on different aspects of the voice: tone, color, harmonics, breath. For rock recordings, intelligibility may be way down the list of qualities an engineer chooses to emphasize as she shapes the voice that will ultimately reach our ears.

What does Siri make of Mick Jagger, I wonder?

Sinatra's Secret

Music with unintelligible lyrics is an object lesson in our ability to detect feeling in the voice regardless of language. And microphones are excellent for capturing precisely that—even better than they do words, as anyone who has struggled to record lyrics with plosives ("p" or "b" sounds) and sibilants ("s" sounds) knows. Great microphone singers like Frank Sinatra maximize the mic's ability to register minute differences in our voices. Every sigh, every breath, every silence can be *felt* via mic. [*Figure 3.9*]

A 1965 CBS television documentary captured Sinatra in the studio recording what became one of his most recognizable performances, "It Was a Very Good Year," together with the exceptional arranger Gordon Jenkins.[6] In it we see Sinatra's superior microphone technique—carefully adjusting his distance from the mic according to phrasing and delivery, making maximum use of what is known as "proximity effect."

Proximity effect is the audio engineer's name for the simple fact that the closer a sound source is to a mic, the mellower its tone; and the farther a sound source, the thinner its sound. Some mics—especially the large diaphragm condenser mics typically used in the studio for recording vocals—emphasize this difference. Others—the small diaphragm dynamic mics most often used on stage for live vocals, for example—minimize it.

FIGURE 3.9

Sinatra during a recording session in a studio at Capitol Records (1953).

Sinatra manipulates the proximity effect by backing away from the mic for those passages he wants to register as louder and more forceful, and leaning in for moments of greater intimacy. His breath control—which Sinatra said he learned from watching Tommy Dorsey play trombone when he worked under him in the 1940s—is justly famous. It is exemplified not only by stamina (in the CBS footage, Sinatra himself is surprised how long he has managed to stretch out the song) but by his ability to avoid sounds from the intake of breath, which would be exaggerated on the sensitive mic he is using.

Sibilants and plosives are similarly held to a minimum by Sinatra's control; in the same documentary, you can hear him discussing this with the engineer. And in a voice-over he provided, you hear his pride at being able to enunciate clearly despite these technical challenges—unlike those young longhairs of the time.

As Sinatra once explained, "You must know when to move away from the mic, and when to move back into it. To me, there's no worse sound than when a singer breathes in sharply, and you hear the gasp over the microphone. The whole secret is getting the air in the corner of your mouth, and using the microphone properly."[7]

U 47

Most people have never been the vocalist on a session using analog recording equipment like Sinatra's. Yet everyone is now familiar with the digital version via the cell phone.

Apple's Siri icon is, ironically, a representation of a classic type of large diaphragm vocal mic: the tube-powered "bottle mic" developed during World War II by Nazi engineers, manufactured after the war in numerous variations but for audio engineers most iconically by Neumann as the U 47.

FIGURE 3.10
Siri icon from Apple iPhone.

Were the iPhone to incorporate such a mic, we would transmit a wall of breath sounds, plosives, sibilants, and ambient noise every time we made a call.

Cell phones are engineered to minimize proximity effect, maintaining a consistency of signal regardless of the speaker's movements. And they are intended to sound the same in any audio environment, whether a crowded bar . . .

FIGURE 3.11
No One Cares (Capitol, 1959).

or an empty street . . .

FIGURE 3.12

In the Wee Small Hours (Capitol, 1955).

Much like navigation by GPS, the cell phone's digital signal processing places the speaker always in the same *non-space*: neither near nor far, neither intimate nor distant. The resulting flatness not only isolates the voice but removes affect. The data is intelligible, but the voice that produced it can only be heard, never felt.

This lack of proximity effect is emblematic of other digital communications. A tweet, a Facebook post, an Instagram photo are all addressed equally to those near and far from us, both in terms of space and relationship. Just as on a cell phone, our ability to modulate tone according to distance is reduced.

Digital media allow for clear communication across great distances, but communicating distance itself becomes a challenge. Online and on the phone, everyone is in the same spatial relationship to everyone else. Proximity effect has been eliminated.

Cell Yell

I began this chapter by whispering to get your attention—precisely what we cannot do on our cell phones.

When we whisper, the noise of our breath competes with the signal of our words, or even overtakes it. Our analog ears, made for pulling sense out of even the least intelligible human sounds, have no problem understanding whispers: not only as speech but as intimate, private speech. The noises of a whisper—the nonlinguistic aspects that make it a whisper—may communicate as much as the verbal signal it contains.

Microphones successfully translate those nonverbal gestures into electrical signals for transmission. When Sinatra leaned into the mic, he leaned closer to our ears. Even an Edison carbon mic transmitted a sense of distance—or intimacy—along the lo-fi, plain old telephone service.

And here we all are, yelling into our cell phones.

"Cell yell," as it has come to be known, is in part the result of where we now use our phones. Engineered as they are to isolate our voices, we routinely employ them in environments where we must shout to be heard . . . were we not on a digital phone. It is also partly due to the fact that we can't hear ourselves when we talk on a cell phone. Digital phones lack what POTS called "sidetone," or what an audio engineer would call a monitor—the sensory feedback of our own voice heard by our own ears. Without sidetone, we all speak as if hard of hearing, yelling into the latest smartphone like Bell did into his original hearing-trumpet version. When we can't hear ourselves, we can't modulate our voices.

But most importantly, we do not hear noise on the cell phone because it's been engineered to transmit only signal. And it's noise—the nonverbal aspects of our voices—that makes the difference between a yell and a whisper. Just as the noise of crosstalk between left and right enables our ears to establish the location of sounds, it's the noisy, nonverbal parts of our voices that establish proximity in our communications. Between our voices and our ears—I'm going to whisper again, this time so you know I really mean it—*we need noise to gauge the distance between us.*

4

SURFACE NOISE

MY FAVORITE RECORDS sound the worst, because I've played them the most. Each time a needle runs around an LP, it digs a little deeper into the grooves and leaves its trace in the form of surface noise. The information on an LP degrades as it is played—as if your eyes blurred this text, just a bit, each time they ran across it.

Analog sound reproduction is tactile. It is, in part, a function of friction: the needle bounces in the groove, the tape drags across a magnetic head. Friction dissipates energy in the form of sound. Meaning: you hear these media being played. Surface noise and tape hiss are not flaws in analog media but artifacts of their use. Even the best engineering, the finest equipment, the "ideal" listening conditions cannot eliminate them. They are the sound of time, measured by the rotation of a record or reel of tape—not unlike the sounds made by the gears of an analog clock.

In this sense, analog sound media resemble our own bodies. As John Cage observed, we bring noise with us wherever we go:

> *For certain engineering purposes, it is desirable to have as silent a situation as possible. Such a room is called an anechoic chamber, its six walls made of special material, a room without echoes. I entered one at Harvard University several years ago and heard two sounds, one high and one low. When I described them to the engineer in charge, he informed me that the high one was my nervous system in operation, the low one my blood in circulation. Until I die there will be sounds.*[1]

Silence is death, the ACT UP slogan painfully reminded us at the height of the AIDS epidemic in 1987. Why seek it out as a part of our musical experience? [*Figure 4.1*]

A–A–D

The switch to digital media for music seems obviously disruptive now, but in the mid-1980s it was so anodyne my musician friends and I hardly took notice. CDs arrived on the consumer market like any other hi-fi marketing scheme, with promises of cleaner sound, greater durability, and a smaller footprint in your living room—all

FIGURE 4.1

Silence=Death poster by ACT UP (1987).

at a correspondingly deluxe price. For those of us happily wallowing in LPs, it sounded like a pitch designed to part bored businessmen from their money. Let them have their new toy, my friends and I thought. Whenever one of our favorite used record stores received a flood of LPs from yet another up-to-date person "converting" their collection, we congratulated one another on our good sense and helped ourselves to more mint-minus albums at rapidly falling prices.

Rumors and conspiracy theories about the CD abounded. "There's no way to permanently bind metal to plastic," a friend who majored in the sciences told me authoritatively. "They'll separate like Oreo cookies." "You know they only cost pennies to make," said a record store clerk we considered a paranoid hippie because he was a few years older than us. "And if you look directly at the red light in the player, you'll go blind." Those who had actually heard a CD play—which wasn't many in my circle, due to the high bar of buying both a new machine and the expensive individual discs—knowingly said they sounded "cold" or "harsh." Hi-fi salesmen explained that the dynamic range available to CDs was greater than our cheap stereos could accommodate; you really had to hear them on an entirely upgraded system to appreciate the difference.

So when my bandmate at the time announced he had bought a CD player in order to hear one of our favorite albums—the Feelies' *Crazy Rhythms*—"without the scratches," I received the news with more than a bit of disdain. And then I eagerly asked to hear it too.

It was true. There were no scratches.

The sensation of first hearing a CD of a recording I had memorized—together with the surface noises on my copy of the LP, and in this case also the (different) surface noises on my bandmate's copy—was something like driving a late-model car designed for a smooth ride rather than my rusting Fiat 128, which had a hole in the floor and struggled to reach highway speed. Just as in a big new American car, I could no longer feel the surface.

FIGURE 4.2

Detail from Harry Smith's handbook of liner notes to the *Anthology of American Folk Music* (Folkways Records 1952).

Despite my bandmate's capitulation and the evident truth underlying at least some of the marketing claims for the format, together we continued to make fun of its high-tech, sci-fi image: small silver discs manufactured in "clean rooms" and played with light. We wrote snide liner notes for the first CD to appear with our own music, a European-only release on a small label from Benelux appropriately named, it seemed, Schemer:

On Saturn they're only just picking up the signal. And the bartender says: these guys have a sound.

The sound is today. A few light years away, but it's still now. Flying out of the mystery and back in your life. By laser beam.[2]

Although we had agonized over every aspect of our first LP, this first CD we treated flippantly—improvising the silly liner notes together on a typewriter, blithely adding a "bonus track" that had previously been hotly debated and dropped from the LP. We were like those Hollywood stars who protected their image fiercely in the American media but consented to embarrassing commercial endorsements in Japan. A CD felt so remote to our lives in music it might as well have been intended not just for overseas but for another planet, as we teased in the liner notes.

The joke was on us, obviously. Precisely what seemed most absurd to us at first about CDs—that nothing need touch them as they played—is what made them truly different from LPs and what ultimately ended the musical era we had grown up in. "Digital" was Orwellian in its misdirection: these were objects nobody handled. By contrast, we put our fingers all over LPs. A friend who owns a record store tells me some collectors even lick them.

Sleight of Hand

Of course, the music industry loved CDs. No one could fold, mutilate, or lick the metallic discs sealed inside their plastic "jewel cases," eliminating the damaged paper sleeves and scratched vinyl that had to be accepted back for credit from retail. What's more, consumers accepted prices that allowed a much greater profit margin than LPs. And artists miraculously accepted lower royalties for CDs, thanks to larger "packaging deductions" labels claimed against what they said were much greater manufacturing costs.

On top of it all, the labels could resell everyone the same recordings they already owned. What was not to love? The music industry was in ecstasy, kissing itself congratulations like Zero Mostel in *The Producers*.

But if the arrival of CDs really were a movie, I would score the first ominous chord to the scene where a shoplifter discovers how much easier it is to conceal these smaller discs.

Not that shoplifting in record stores was new. Theft is as old as property, and in truth anything can be stolen from anywhere, with practice (see Bresson's *Pickpocket*). Yet CDs seemed to send retail into a panic, and next we knew the bag-check station at the front of larger record stores—along with their intimidating attendants—had been removed in favor of what looked like the metal detectors familiar then only from airports and courthouses.

Were those bag-check jobs the first to go?

The newfangled antitheft systems installed for protecting CDs were "Electronic Article Surveillance" (EAS)—the now ubiquitous practice of tagging merchandise with metal strips, deactivated with a pad at the register. (Which is actually magnetizing them; it is the irregular atoms of an "amorphous" metal that triggers the system.) The five-finger discount was countered by disembodied surveillance.

These new surveillance systems had an immediate effect on the culture of record stores. Employees no longer had to watch while you browsed to make sure you weren't pocketing goods. And there was less reason for them to engage you in conversation—your motivations for being in the store, your knowledge of their specific merchandise, your identity as a local or transient, all became insignificant for the detection of potential theft now that there was an all-seeing machine by the door.

The difference between shopping for music in larger and smaller stores quickly grew more exaggerated. Smaller shops—lacking both the capital and floorspace to take full advantage of the increase in catalog sales fueled by the CD boom—remained more or less unchanged, still leaning toward LPs and their specialized knowledge of a portion of the market. In Cambridge we were blessed with a wealth of these shops: Stereo Jack's, for jazz; Skippy White's, for soul and R&B; Newbury Comics, for hardcore and what was then called "college rock"; Festoons, for U.K. and European imports; Second Coming, for bootlegs; Cheapo, for cutouts. Not to mention the used record stores: Mystery Train, In Your Ear, Looney Tunes . . .

Each of these smaller stores was watched over by a staff dressed to match the genre of records they preferred, and/or the owner in whose image the place had been made. "Get out of my light," I once heard Stereo Jack say to a customer. He was busy looking for scratches on vinyl and couldn't care less what she might be doing there (she was shopping).

The biggest, most comprehensive record store in town before the introduction of CDs was the Harvard Coop, which also supplied students with textbooks, desk lamps, and standard-issue preppy clothing. But once the massive profits of CDs entered the picture, so did national and even transnational chains. Suddenly Cambridge had a branch of the legendary California store Tower Records and, almost directly across the street, an even bigger HMV from the U.K. Just over the river in Boston, strategically placed on the Mass Ave bus route that runs between Harvard, MIT, and the Berklee College of Music, Tower opened the biggest store of all in a turn-of-the-century warehouse, dramatically expanded and reconstructed by Frank Gehry in 1989. Gehry's Tower Records building was as visible from the Mass Pike, and for a time nearly as iconic, as nearby Fenway Park. [*Figure 4.3*]

These megastores were less record shops than retail outlets. They had rows of registers staffed by disaffected employees. They had pricing wars, undercutting both the local stores and one another.

FIGURE 4.3

Frank Gehry's Tower Records Building, Boston (1989). Tower retail occupied five full floors—most of the building has since been converted to luxury condos.

They had celebrities from MTV making personal appearances. And they were jammed. I shopped at them, too—everyone did because the discounts were so deep. Sometimes I'd see employees from the smaller record stores in there, buying.

Today, both the Tower and HMV chains are bust. Here in Cambridge, the former Tower Records is a vast Verizon Wireless showroom. And the huge HMV space across the street sits empty—it's been dark for a decade, in the middle of Harvard Square. It seems no local business can afford the rent on a space that large.

Welcome to the Digital Music Revolution

Once music was digitized, it was only a matter of time (Moore's Law again) before it became more quickly and more cheaply available. The CD did, in a way, prove to be as absurd a format as it first appeared—why use mini LP-shaped discs for digital files? Why use any physical object to distribute digital information?

Napster in its original, free-for-all form existed only two years, 1999–2001, but once it had liberated music from those small shiny discs there was no sticking it back. Apple launched iTunes in 2001 and the iTunes Store in 2003, capturing the online exchange of digital music files for profit and ultimately killing the CD altogether. "Welcome to the digital music revolution," said Apple's ubiquitous ad campaign for its new digital music player, the iPod.

The iPod eliminated any physical embodiment of the individual sounds played on it. Not only were there no more discs, large or small, but there was no more packaging. There was nothing left to touch except Apple's "click wheel"—a last vestige, perhaps, of the idea that records go 'round. [*Figure 4.4*]

The Pianolist

This intangibility of digital music has a precedent from the earliest days of sound recording.

Before the Victrola and the radio, music in the home meant instruments in the parlor. ("Parlor guitar" remains the term for a small-bodied acoustic.) The piano was—still is—the grandest, most expensive, and least portable parlor instrument of all. In the United States, the post–Civil War economic boom was marked by a flood of pianos. To this day many occupy a central place in older American homes, whether or not anyone plays them or even wants them; the website PianoAdoption.com maintains a list of those available for free to anyone willing to move one. (It was developed by a very clever piano mover from Nashua, New Hampshire.)

All those pianos needed sheet music. As early as the 1830s, Boston composer and churchman Lowell Mason (his settings of hymns are still familiar to many Americans) advocated that music be taught in the newly developed public school system. By the time

FIGURE 4.4
Clickwheel from the iPod Classic.

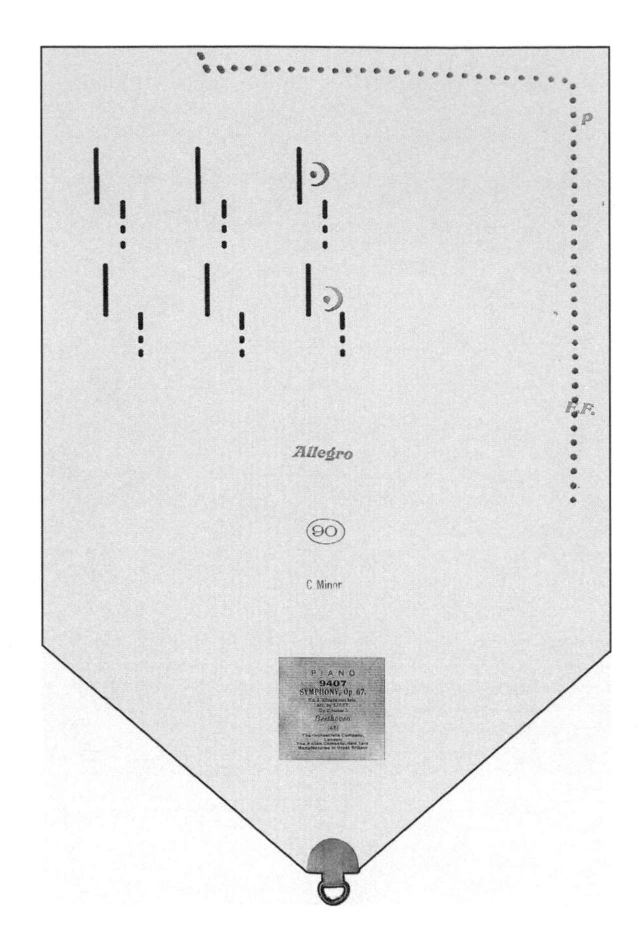

FIGURE 4.5

The opening of Beethoven's Fifth on a roll for a player piano. Non-perforated marks are intended to be read by a "pianolist," the operator of the machine's foot pedals who was responsible for tempo but would never touch the instrument's keys.

Lowell's son Henry started manufacturing Mason & Hamlin pianos in the 1880s, "America had become the most musically literate nation on earth," according to the Center for Popular Music.[3] In the Gilded Age, music publishers were as formidable a presence for U.S. intellectual property as piano manufacturers were for the industrial economy.

Then in 1898 a disruptive digital invention pitted one against the other: the Pianola or, as it came to be known generically, the player piano.

The player piano dispensed with the need for sheet music in favor of a piano roll directing air-powered levers. The piano roll is a pre-electronic digital technology—like the Jacquard loom, it uses punches in paper for "on" and "off" binary instructions. The first device to make use of this technology, the Aeolian Company's Pianola, proved so popular that by the 1920s half the pianos sold in the country had incorporated it. Even Steinway was making player pianos.

While piano manufacturers might benefit from this new technology, sheet music publishers could not. The technology of the piano roll belonged exclusively to its makers. And by 1902, only four years after the launch of the Pianola, they were selling more than a million of them. [*Figure 4.5*]

So the sheet music publishers did what any software company would do when a hardware manufacturer threatens to make its product obsolete: they sued. The publishers argued that the digital piano roll violated their copyrights by reproducing the music they printed, even if it didn't make use of their product to do so. Their case went all the way to the U.S. Supreme Court. And they lost.

In 1908, the Supreme Court ruled in favor of a Chicago manufacturer of player pianos and piano rolls and against a Boston music publisher who had sued over use of the songs "Little Cotton Dolly" and "Kentucky Babe." The Court reasoned, in *White-Smith Music Pub. Co. v. Apollo Co.*, that music is not a "tangible thing": "In no sense can musical sounds which reach us through the sense of hearing be said to be copies," wrote Justice William R. Day for the majority, reasoning that they were therefore not subject to *copyright*.

> *A musical composition is an intellectual creation which first exists in the mind of the composer; he may play it for the first time upon an instrument. It is not susceptible of being copied until it has been put in a form which others can see and read.*

Piano rolls can be seen, to be sure, and it might even be said that they can be read—but not as music, or at least not by a person. Therefore the Apollo Co. could continue to manufacture piano rolls of "Little Cotton Dolly" and "Kentucky Babe" with impunity, said the Court, since "these perforated rolls are parts of a machine."

The Court went on to note that this same reasoning would apply to another recent invention: the wax cylinder recording. Here

Justice Day approvingly cites language already used by the Court of Appeals:

> *It is not pretended that the marks upon the wax cylinders can be made out by the eye or that they can be utilized in any other way than as parts of the mechanism of the phonograph. Conveying no meaning, then, to the eye of even an expert musician, and wholly incapable of use save in and as a part of a machine specially adapted to make them give up the records which they contain, these prepared wax cylinders can neither substitute the copyrighted sheets of music nor serve any purpose which is within their scope.*

The decision left music publishers empty-handed, as it were. Sound wasn't a tangible thing, and their copyrights were tactile only. The manufacturers of the new Pianolas and Victrolas owned the patents to all the mechanical parts of their devices, and if those parts emanated music that was their business.

This didn't go over well in Congress. The next year, it rewrote copyright law to supersede the Supreme Court's ruling. Looking to rescue music publishers from the Napster-like chaos of royalty-free piano rolls, yet allow the player piano industry to continue manufacturing without being hamstrung by intellectual property owners, the Copyright Act of 1909 established a system of compulsory mechanical licenses. Mechanical reproduction of music (i.e., piano rolls, gramophone records) could continue without permission of the music publishers, so long as those publishers were paid a statutory royalty for each "mechanical reproduction" derived from use of their music. (Songwriting royalties are still calculated this way and are known as "mechanical royalties.")

However, Congress declined in 1909 to redefine what constituted music, allowing it to remain in the eyes of the law an intangible thing. Surprising as it may seem in retrospect, "musical sounds which reach us through the sense of hearing"—recordings—remained outside U.S. federal copyright protection until February 15, 1972. Which explains the twentieth-century music industry's focus on the "label"—a tangible and therefore copyrightable object that took on such outsize legal importance it became a metonymy for the record company itself. Since sound could not carry copyright, the © ownership symbol on record labels and sleeves applied only to what was printed on them: logos, artwork, liner notes.

In 1972, the U.S. law was amended to allow for copyright of sound recordings and a separate ownership symbol was established because © hadn't previously applied: ℗, for phonogram.

Mind Your ℗s and ©s

When Napster gave the lie to CDs, making clear that the digital sounds they contained were unbound to silver discs, it reawakened the early twentieth-century dispute about player pianos. Once again, the owners of intellectual copyright—the record "labels"—came into conflict with developing digital technology, in this case the personal computer. Computers could swallow those expensive printed discs, strip the digital music files off them, and spit them back out as worthless plastic.

As before, the owners of copyright went to court—but this time they won. Napster's peer-to-peer network for sharing digital sound files went live in June 1999, and by that December they had been sued by a consortium of major labels represented by the Recording Industry Association of America (RIAA). The suit took two years to work its way through the courts before an injunction was issued against Napster. These were the only two years Napster existed in its original form.

The RIAA's lawsuit tried to stop peer-to-peer sharing of files by shutting down Napster. But regardless of whatever happened to Napster, it couldn't reglue digital files to a physical format. With iTunes, Apple stepped into the space Napster had opened up between the disc and the sound files, supplanting the functions not only of the record companies—pressing and distribution—but also the retail record stores.

Apple left out a part of the old puzzle, however. All that information record companies print on sleeves and labels—the only information they could protect under federal copyright before 1972—never made the transition into the iTunes Store, which attaches merely title and artist (alphabetized by first name rather than last, no less) to its files. Songwriters, producers, engineers, publishers, even the musicians on recordings are rendered anonymous. In other words, Apple took the phonogram, ℗—precisely what the Supreme Court said in 1908 could not be owned—and ignored the tangible printed matter, ©.

Which is perhaps the true digital music revolution, from the piano roll to streaming files. Digital music is all ℗: sound you cannot touch.

Sticky Fingers

Let's return for a moment to the anechoic chamber with John Cage.

Cage entered that room to experience silence—what audio engineers now call digital black, the absence of both signal and noise. But his own living body, he discovered, emitted sounds in time: the sounds of his nervous system in operation, and his blood in circulation. "One need not fear for the future of music," concluded Cage—because what is music but sounds in time?

Silence is beyond our corporeal experience, since living bodies occupy not only space (the anechoic chamber) but time (John Cage *in* the anechoic chamber). We can imagine and create the conditions for noncorporeal sound, but we cannot experience it because we hear in time. Our ears are as sticky as our fingers. And what they stick to is *time*.

This is what makes the Supreme Court's decision of 1908 intuitively wrong, regardless of one's legal judgment. Sound in the abstract may not be a "tangible thing," as the Court asserted, but *sounds in time* are. The invention of audio recording made this clear to people immediately. "Canned music"—John Philip Sousa's term for recordings when they first appeared[4]—is music stored for the future. It is bottled time.

As Jonathan Sterne details in his history of early recording, the invention of canned music was not unrelated to a contemporary fascination with embalming.[5] The Victorians were death-obsessed and saw sound recording as another means of preservation: "Death and the invocations of 'voices of the dead' were everywhere in writings about sound recording in the late nineteenth and early twentieth centuries," Sterne writes. He points out that even Nipper, famous mascot and logo for HMV ("His Master's Voice"), is based on a painting of a dog listening to a gramophone that many assumed was placed on top of a coffin. [*Figure 4.6*]

Nipper responds to the recording of his late master's voice because the sounds it reproduces are tangible in time. It's just that time has been displaced.

The Splendid Splice

The invention of magnetic audiotape in the 1940s made this displacement of time literally more plastic. While wax cylinders and gramophone records could preserve a solid slab of time, tape could be cut into pieces of time and rearranged. Glenn Gould called this

FIGURE 4.6

Francis Barraud, *Dog Looking at and Listening to a Phonograph* (1898–99). This photograph of Barraud's original painting was filed alongside a "Memorandum of Assignment of Copyright" with the Public Records Office in London. It featured an Edison cylinder recording device sold primarily for use as a dictaphone. Later, Barraud altered the painting to represent a Gramophone and retitled it *His Master's Voice*. Nipper was Barraud's late brother's dog.

"the splendid splice," because it allowed him to perfect a recorded performance by picking and choosing among parts of different takes.[6] A razor blade and some sticky tape was all it took to join one moment in time to another. [*Figure 4.7*]

Experimental composers quickly pushed this plasticity to an extreme in the pursuit of abstraction. "Musique concrète," as formulated by French composer and theorist Pierre Schaeffer, used the splice to sever the "sound object" from its source (an instrument or a field recording location), which might then be rendered unrecognizable by abbreviation or other manipulation. John Cage used the splice to reorder sounds according to chance operations—although the immense labor required to turn the 192-page score of his first tape piece, *Williams Mix* (1952), into the resulting four and a half minutes of music dissuaded him from pursuing the technique much further. Each page of Cage's score, which specifies multiple splices in two "systems" of eight tracks of tape each, sums to just one and one-third seconds of playback. [*Figure 4.8*]

One might assume that the dense number of splices in a work like *Williams Mix* would lead to nothing but a blur of undifferentiated noise. Yet even in such an extreme work, where more than five hundred source sounds have been cut and shuffled in fine detail, there is an unmistakable recognition of *sounds in time.* Our ears catch extraordinarily small moments as they rush by, whether in recorded music or in the world.

Audio engineers have tested the limits of this perception by looking for the shortest duration of sound we can recognize as a note. The answer is 100 milliseconds. In *Microsound*, Curtis Roads reports that in even less than that amount of time our ears can still perceive "discrete events . . . down to durations as short as 1 ms."[7] Those are heard as clicks—but clicks with "amplitude, timbre, and spatial position," which can therefore be distinguished from one another.

A millisecond, in case you aren't familiar with that end of the timescale, is one thousandth of a second. Imagine Cage's score for *Williams Mix* stretching to 192,000 pages for the same four and a half minutes of sound. No analog work could begin to approach that level of detail.

Or we might say: no analog work can exceed our powers of perception for time.

The Foothills of the Headlands

In popular music, tape manipulation pushed toward a different set of conclusions, more super- or surreal than abstract. Even before the advent of multitrack machines, artists and audio engineers realized they could "bounce" between two tape decks, overdubbing additional sounds on top of what had previously been recorded. Four-track tape made this process flexible and efficient enough for

FIGURE 4.7

Electronic music composer Delia Derbyshire splices tape at the BBC Radiophonic Workshop (1965).

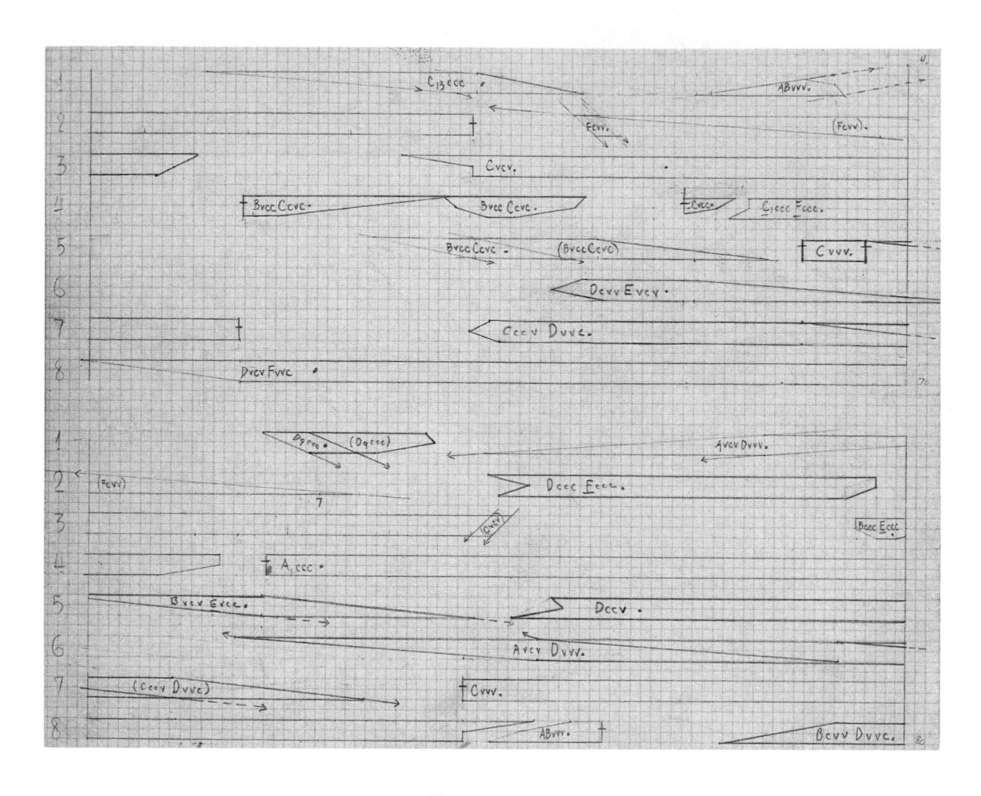

FIGURE 4.8

One of the 192 pages from John Cage's *Williams Mix* (1953). Each line is a track of tape "pictured full-size, so that the score constitutes a pattern for the cutting of tape and its splicing."

the Beatles to record their psychedelic masterpieces *Revolver* and *Sgt. Pepper's Lonely Hearts Club Band.* After filling all available space, Abbey Road engineers would make a "reduction mix" to a single track (either on the same tape or to one on a second machine), and continue adding on top.

Overdubs make different use of the time embodied on magnetic tape than a splice. While a splice joins one discrete moment to another, overdubs layer multiple moments atop one another to make a super-real environment—one in which string orchestras and backwards guitars move together through the same space of time, on a single piece of tape unspooling at fifteen inches per second.

Listeners to these imagined soundscapes seized on their hyperreality rather than their impossibility. "Lucy in the Sky with Diamonds" is an archetypal song of the era, perhaps not only for the implicit drug reference (which singer/songwriter John Lennon always denied) but because it describes what it's like to hear a multitrack recording. "Picture yourself in a boat on a river," it begins, as you might have done while listening to Debussy. But it then adds an unforeseen layer of color: "with tangerine trees and marmalade skies." As you adjust to this synesthesia and begin to focus on "a girl with kaleidoscope eyes," Lennon's voice suddenly moves much farther away, singing:

> *Cellophane flowers of yellow and green,*
> *Towering over your head.*

Evidently, it may be you who has moved by growing very small; Lennon's voice might well have remained where it was. But where does that put the girl we were just getting to know?

> *Look for the girl with the sun in her eyes,*
> *And she's gone.*

Boom—boom—boom. Not only the girl but everything in the soundscape disappears, cleared away for a chorus that emerges in an entirely different space again. A space you enter too, as inevitably as one moment follows another.

John Lennon pulls us through the shifting perspectives of "Lucy in the Sky with Diamonds" as if guiding us through the multiple layers of time and space the Beatles added to their multitrack tapes. Like John Cage's 192-page score for four and a half minutes of music, each of the brief pop songs on *Sgt. Pepper's* represents hundreds of hours of labor. But rather than compressing that time by cutting it up as Cage had, the Beatles layered over and over the same length of tape, until it was so thick with time that listening to it reminded people of an acid trip.

Tape Hiss

Just as there is a physical limit to the number of splices that might occupy a given length of tape—a limit John Cage seemed to approach on his very first pass, in *Williams Mix*—there is a limit to the number of overdubs possible in an analog medium. Tape itself is not silent as it moves through a recording machine; no more than we are in an anechoic chamber. Which means each overdub adds not only more signal but more noise, in the form of tape hiss. And layers of hiss don't get more trippy, they just get louder.

One reason the great works of multitrack analog recording were made by artists with tremendous resources—the Beatles, the Beach Boys—is that it took the finest analog equipment to keep tape hiss at a minimum for that many passes through the machines. "Lo-fi" artists have made equally dense and psychedelic works; the Elephant 6 collective of the 1990s were still in high school when they started making theirs, on four-track cassette. But in analog recording, overdubs and tape hiss necessarily go hand in hand; only capital (or Capitol) can keep the latter manageable as the former pile up.

Even so, *Sgt. Pepper's* and *Pet Sounds* are works of noise as well as signal. Those noises are not limited to tape hiss—they include all the many aural artifacts of the various times and spaces layered onto these short lengths of tape. A well-known example on *Sgt. Pepper's* is the studio air-conditioning audible at the end of the album's dramatic final chord. And obsessives have made use of internet crowdsourcing to catalog all the many noises married to the signals on Beach Boys recordings.[8] Here is the list just for the song "Here Today," from *Pet Sounds*:

1:15 Mike starts singing the chorus too soon. "She made me feel" and then someone else says something to make Mike stop

1:27 Something metallic is dropped at the point: "She made my heart feel sad. Sh(drop)e made my days go wrong . . ." of the second chorus.

1:46 Brian says "Top" as soon as the second chorus ends to rewind the tapes and start the take over

1:52 Someone says something supposedly about cameras

1:56 Someone else replies to the person at 1:52

2:03 Brian says "Top please," probably because he realizes the tape is still rolling after all these noises

2:20 Talking

These inadvertent noises are inseparable from the intended signals on the tape. Had Brian Wilson wanted to get rid of them, his only option would have been to rerecord the entire track on which they occurred. Had that track already been bounced along with others in a reduction mix, it would mean rerecording all those tracks too. And had the unintended sounds gone undetected until the final mix, as often happens, it would mean throwing away the complete recording and starting all over again.

Analog recording is an *additive* process. Whatever happened in the studio as each layer was added, happens again on the tape as it unspools. For all the Abbey Road engineers' ingenuity—which was truly remarkable, they seem to have utilized or invented most every analog studio recording technique—they could not remove the air-conditioning at the end of "A Day in the Life" without removing the dying piano chord as well.

Thick Listening

At the other end of that additive process is the close listener. If you listen closely enough to an analog recording, you hear all its sounds preserved together: the signal and the noise.

When the catalogers of unintended noises listen to Beach Boys records, they listen between the notes. We might call it *thick listening,* alert to the depth of the many layers in multitrack recording. They listen through the surface noise of the LP, through the hiss of the master tape, through the layers of the music itself all the way back to the room in which it was played, where two horn players are standing and chatting.

In other words, they are listening to more than the signal of the music—they are listening to the signal *framed and enriched by noise.*

Do digital formats reward this kind of attention? Our developing habits would seem to indicate otherwise.

In iTunes, I keep a folder of music I have access to only in digital format—mostly bootlegs found on the internet and promo copies sent to me as downloads. I don't think of it as a large part of my music library, since I own more records and CDs than are reasonable for apartment living. Yet iTunes calculates that it would take me five days, fifteen hours, fifty-one minutes, and five seconds to listen to it all.

Will I ever?

Frictionless digital music—those sounds we cannot touch—is distributed and stored without friction as well. Apple's iPod Classic was touted for its ability to hold up to 40,000 songs. For scale, the Beatles wrote a total of 237 songs.

It is normal, with today's digital media and devices, to have access to far more music than one can ever hear. The time it takes to listen to music is now in shorter supply than recordings. Digital music has created a time deficit.

Which means that even in my relatively small folder of digital bootlegs and promos, many will likely go unheard. More to the point: most will never be listened to *closely.* Close listening is a function of time. It starts at the beginning, takes in each note and the spaces between, and stops at the end.

Does that describe our digital listening habits? I for one find myself clicking through a good deal of digital music. If it's online or on my computer, I skip around—I preview tracks, hearing a bit here, a bit there. My digital listening is to *signal alone.* I hear the notes but not the space between, or the depth below. It's listening to the surface without the noise.

5

LOUDNESS WARS

IN THE 1990s, RECORDED MUSIC GOT LOUDER. A lot louder. "From the mid-1980s to now, the average loudness of CDs increased by a factor of 10," explained a 2007 article in the journal of the Institute of Electrical and Electronics Engineers (IEEE).[1] The next year these "loudness wars" hit a peak, as it were, with the 2008 release of Metallica's much-maligned album *Death Magnetic*. Even for Metallica fans there was such a thing as too loud—thousands of them signed a petition, "Re-mix or Remaster Death Magnetic!" which proclaimed, rather kindly: "*Death Magnetic* is a fantastic effort from Metallica, however many of us are disappointed by the poor audio quality present on the album."[2]

Metallica fans knew a better-sounding release was possible because a different version of the same album had surfaced as part of the video game *Guitar Hero*. The video-game version was identical, just *less loud*. And it turned out you could hear more of the music that way.

Audio engineers, including Matt McGlynn in *Recording Hacks* and Paul Abbott in *Tape Op*, were quick to explain what had happened. The video-game makers had evidently been given the same mixes, but without the memo to "brickwall" their mastering in order to maximize volume. The CD master was much, much louder—and had sacrificed all dynamics to get there. Putting the two versions of the Metallica album into a digital audio workstation (DAW) program provides a graphic illustration of the difference: the CD is a solid block of sound, while the video-game version has peaks and valleys marking louder and quieter material. [*Figure 5.1*]

The band was unmoved by the petition. "I've been listening to it the last couple of days in my car, and it sounds fuckin' smokin'," Lars Ulrich told *Blender* magazine when asked about the controversy.[3] Of course Lars was likely driving a very fast and thus very loud car—not the best environment for evaluating audio quality, though perhaps a kind of ideal for listening to Metallica. "I don't know what kind of stereos these people listen on," he sniffed.

There's an aspect to the loudness wars reminiscent of Spinal Tap's Nigel Tufnel turning up to eleven ("It's one louder, innit?"). And life imitated art nowhere more closely than on Oasis's massive, and massively loud, 1995 hit *(What's the Story) Morning Glory?* which many see as the Sputnik launch of the volume race. Producer Owen Morris—pictured in the background of Oasis's album cover, face hidden by a box of recording tape—later explained

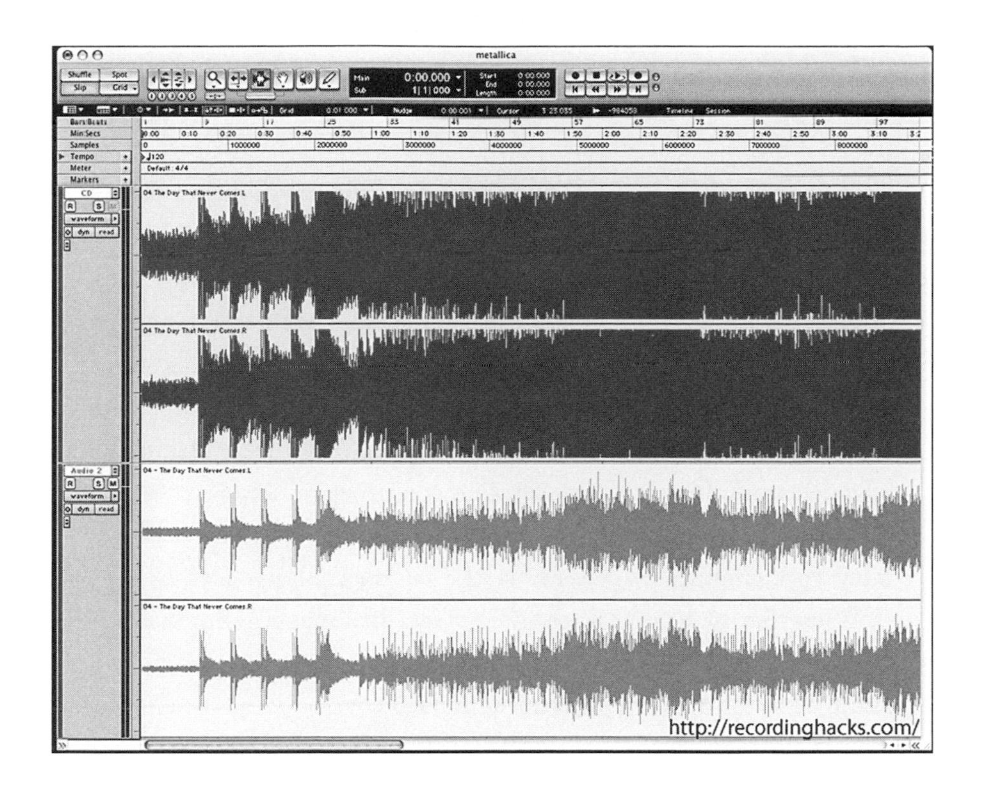

FIGURE 5.1

At top, the beginning of a song from the CD *Death Magnetic* by Metallica; below, the same excerpt from the version embedded in the video game *Guitar Hero*.

in a kind of humble brag: "Given that I had no confidence in the sonic integrity of my mixes I had decided that I would attempt to use VOLUME (i.e. quantity rather than quality!) as my rather blunt tool."[4] Mastering engineer Barry Grint remembers Morris showing up with a Digital Audio Tape (DAT) of his final mixes. Pointing to the peaking meter lights as it played, he instructed Grint: "See how the top three LEDs only light up sometimes? By the time you've finished, I want them lit up ALL the time!"[5]

Problem was, those lights were already indicating a maximum of digital information on the DAT—and Morris wanted the master even louder. The resulting CD was derided by audio engineers, yet it proved so popular that Morris now claims full credit for its overloud mastering as well as its blunt-tool mixes. As for Grint, he seems to have felt from the start the mastering was an error, and remembers that the U.S. record label objected to it on technical grounds—for a New York minute. "Two weeks later, they were all doing it."[6]

What they were all doing—what Metallica was still doing a decade later, on *Death Magnetic*—was taking advantage of digital audio's lack of noise to push its signal as far as possible. In analog media, raising the signal necessarily means raising the noise level as well. But if there's no noise on a DAT or CD, figured Morris, why hold back the signal? Like a cell phone conversation, Oasis's hits were legible no matter where you heard them—in a bar, out a car window, while shopping at the mall—because they were *all signal*. Every bit of their music came screaming out of the speakers at the same volume as all the rest of it. There were no soft parts to miss, and there was nothing to listen closely for. You got it all, always, whether you wanted it or not.

A recording without dynamics, it turned out, sounded huge and sold correspondingly. *(What's the Story) Morning Glory?* went fourteen times platinum in the U.K. Even Metallica's *Death Magnetic*, despite all the controversy, hit number one on the *Billboard* charts. And its producer, Rick Rubin, won the Grammy for Producer of the Year.

"If we need that extra push over the cliff, you know what we do?" explained Nigel Tufnel. "Eleven. Exactly. One louder." [*Figure 5.2*]

Everything Louder Than Everything Else

Oasis may have initiated the loudness war, but they weren't the first band to push themselves over a cliff. "Can I have everything louder than everything else?" is how Deep Purple's Ian Gillan put it in 1972, asking for adjustment to his stage monitors while being captured for posterity on the live album *Made in Japan*. At the time, Deep Purple were listed by the *Guinness Book of World Records* as the loudest band on earth—a title they would soon relinquish to the Who, with their stacks of Hiwatt amps.

And then along came Kiss, with entire walls of Marshall amps. Although, what's that one microphone doing in front of just that one speaker. . . ? [*Figure 5.3*]

"I'm not gonna bullshit you," Kiss guitarist Ace Frehley told

Fuzz magazine when he was on a reunion tour in the 1990s and perhaps a bit less invested in image than he had been twenty years earlier. "Did you honestly think that all those speakers are working?"[7] Kiss's empty Marshall cabinets were an open secret in the music industry, much admired and copied for their stagemanship. (A more difficult rumor to confirm is that Kiss actually played through Fender amps kept entirely out of view.)

Kiss played through no more amps than any other band because more volume doesn't necessarily add up to more loudness. In fact it's difficult to use multiple amps to effect, because sound waves out of phase cancel one another out—the same principle used by noise-cancelling headphones. Regardless of a given guitarist's amp technique, for large venues like those where Kiss appeared, the sound heard by the audience is not from the amps on stage at all but from the public address (PA) system. Those speakers—the large, multiple arrays placed between stage and audience—have been "tuned" for phase and equalization in order to work together rather than in opposition, and they have a maximum output of their own. Putting more sound from more amps through the PA doesn't add to its potential volume any more than painting a number 11 on a dial that previously read 10.

FIGURE 5.2

Nigel Tufnel's custom Marshall amp, frame from *This Is Spinal Tap* (1984).

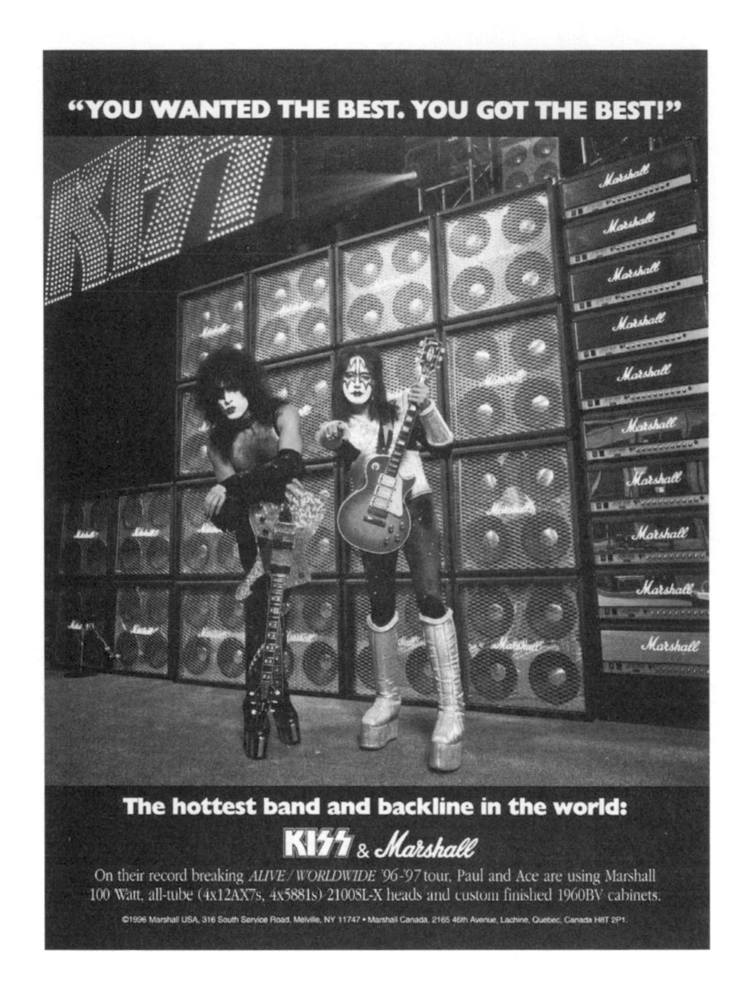

FIGURE 5.3

Marshall ad for Kiss endorsement of their brand, from *Guitar World* (January 1997). Stage right is the one speaker in the one cabinet mic'd for actual amplification.

Into the Red

In the analog studio, recording tape functions like the PA system of a venue; the listener, in the end, doesn't hear the volume of the amp being recorded but the volume of the tape. And tape, too, has a maximum output.

So what happens when you push the tape to 11? The answer is the sound of most every rock record from the analog era.

Analog media are *nonlinear*: as you increase the amount of information directed to recording tape, there is not an equivalent increase of information stored. Instead, the medium imparts a curve to that information, determined by its own qualities of responsiveness. It's like us, in other words—some information we take in, some we change as we absorb it, and some never quite gets through.

Tape has a particular set of characteristics that influence the resulting curve, which audio engineers have labeled with wonderfully evocative names: wow and flutter, self-erasure, hysteresis, saturation. Like Cage in the anechoic chamber, tape in the studio is a presence contributing to sound, not silence. Perhaps because we maintain an idea of transparent recording, if not the fact—just as there is an idea of silence we cannot physically experience—the ways in which an analog medium shapes sound are collectively known as *distortion*.

Which brings us back to the edge of that cliff rock bands rush toward. More volume may not necessarily lead to more loudness, as Nigel Tufnel assumed. But pushing everything to 11 does create more distortion. In the analog studio, sending more signal to tape exaggerates the curves imparted by it. You might say it invites the tape's greater contribution to the recording.

The volume unit (VU) meter is a tool from the analog recording studio that shows up in consumer equipment as well. It is an exceedingly simple (and cheap) device, a needle that moves up and down along with volume. The reaction time of a VU meter is slow enough that it doesn't respond to every tiny fluctuation—which is fine, because it is used to measure averages rather than specific units. The scale it points to is similarly vague; it has a zero point, with a section below in black and a section above in red. The goal in adjusting the volume of a signal for recording to tape is to keep the VU meter's average around zero—which means its needle necessarily spends a good deal of time in the red.

If you were around for the era when home taping killed music, you probably figured out on your own cassette deck what every professional audio engineer knows about tape: it sounds better if you keep the signal hotter, that is, with the needle of the VU meter in the red as much as possible without "pinning" it to the top. Saturating the tape in this way increases the signal-to-noise ratio—meaning you hear more music and less tape hiss when it plays back. On cassettes, this was particularly clear because their inherent hiss was so loud; the small format and cheap machines they played on could make for more noise than signal if you didn't push the levels far into the red.

Rock music is well suited to this strategy of pushing levels into the red because it can make good use of the distortions introduced by tape. For example, saturation compresses quick, fast-moving peaks of signals into the main body of the program. Think of Ringo Starr's characteristic cymbal sound on Beatles recordings, like the crashes that punctuate "Taxman" at the start of *Revolver*: washes rather than clicks, sitting "in" the mix rather than "on top." That's the effect of tape compression, exaggerated by multiple passes through the machines during overdubs.

The distortions of tape saturation reduce its ideal fidelity, but depending on the program that isn't necessarily a bad thing. Symphonic music doesn't generally benefit from this kind of compression. However, a composition like Terry Riley's *In C*, which leans on textures and layers, may sound only more fantastically complex as tape works its magic.

Metal Machine Music

Magic is exactly what many musicians in the analog era ascribed to the recording studio, not least because of the distortions from tape. Sounds can appear on tape that no one played, whether from the harmonics generated by microfluctuations in speed known as wow and flutter, or from the collision of tones compressed by saturation. And blips of sound that felt harsh or even out of time as they were performed sometimes disappear, fallen into the gaps caused at the louder end of the spectrum by self-erasure, or at the softer end by hysteresis. The tape itself might seem to add that crucial bit of "fairy dust," as it was described during a famous studio argument among the Troggs.[8] (If it didn't, you could always try following the Troggs' advice of adding a twelve-string and/or pissing over the tape.)

The German band Can were particularly devoted to studio magic and approached it with the gravity one might expect of 1970s rock musicians who had studied with Stockhausen. "In a studio you make a concert for machines, and machines really like to listen," explained bass player and producer Holger Czukay. "They have a heart and soul—they are living beings. The Can recordings are best . . . when the desk and equipment are treated like human beings."[9] [*Figure 5.4*]

Guitarist Michael Karoli described a particular session:

> *There was a whole evening when we had been talking about things like recording and afterwards finding something on the tape which nobody heard while it was being recorded, so we decided to do the same thing, to put our equipment in and make music by not doing anything.*[10]

The result of experiments like this were songs such as "Unfinished," the thirteen-minute track that closes Can's 1975 album *Landed*, which keyboardist Irmin Schmidt explained was "a selection of

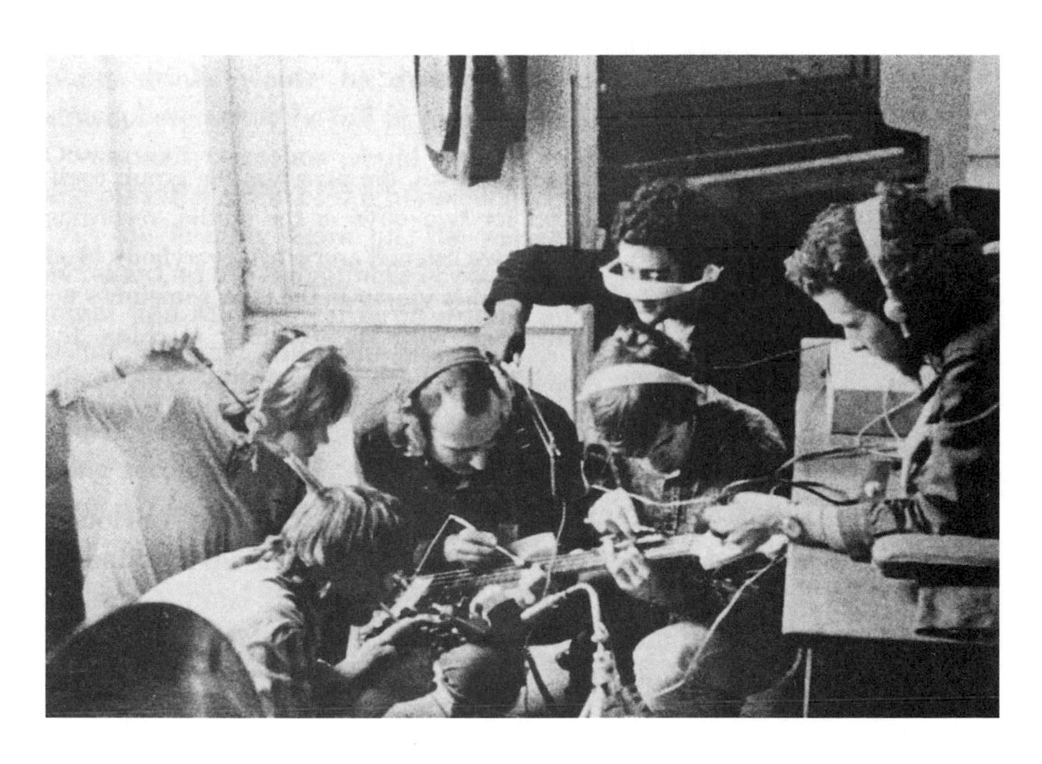

FIGURE 5.4
Can in the studio (1968).

a 45-minute session we called *The Magic Day*, it was a very magical recording." Adds Karoli: "'Unfinished' is one of those pieces where we let the atmosphere of the studio impress itself on the tape, it's really a piece composed by the studio."[11]

When the studio is asked to compose on its own, as it was by Can, it uses noise as signal. "Unfinished" is music comprised of noises—some generated by musicians playing instruments, some generated by the studio playing tape. And true to Holger Czukay's description of machines as living beings, it's difficult to say which is which. In Can's studio technique, noise and signal are equally significant materials, and in a song like "Unfinished" they are equally compelling. Can's machine music is not harsh or repellent to the ear—it's inviting and beautiful. The noises in it are no less human than the signals, just as Cage discovered in the anechoic chamber.

Beautiful Noise

Like many avant-garde experiments, Can's use of the recording studio highlights qualities that are present but usually overlooked in more mainstream work. The studio is *always* a composing partner in recording. As Jonathan Sterne has written about the earliest days of the medium, "Recording is a form of exteriority; it does not preserve a preexisting sonic event as it happens so much as it creates and organizes sonic events for the possibility of preservation."[12] What goes to tape is what the studio makes possible, just as Can's

Michael Karoli said of "Unfinished": "We let the atmosphere of the studio impress itself on the tape."

This "exteriority," as Sterne terms it, is endemic to the act of recording and applies equally to the digital studio. But Can's work highlights a quality of the analog studio that has not carried over to digital. In analog recording, the medium contributes sound alongside the musicians. That's the fairy dust of tape: *noise as part of the signal.*

Indeed, it is the music itself that determines the harmonic distortions introduced by an analog studio. Take wow and flutter, the distortion with a name as swinging as the 1960s ads that promised to minimize it on your hi-fi. The term refers to distortions from imperfections in the speed of playback on analog machines—the wobble of a warped record would be an extreme example. But consider how even a huge wobble is heard as a variation of the program being played. If the warped record is an orchestra playing Mozart, the wobble is in Mozart's musical vocabulary. If it's *Led Zeppelin IV*, it speaks Jimmy Page.

Professional tape machines and expensive hi-fi systems are calibrated to minimize such speed distortions, though they can never eliminate them entirely. What remains of wow and flutter on the best analog equipment—say on the machines Jimmy Page used to record *Led Zeppelin IV*—produces a very subtle effect. At those tiny levels, the speed fluctuations of wow and flutter don't sound like the wobble of a warped record; they sound like vibrato and harmonics. They sound musical.

All distortions of tape are, like wow and flutter, bound to the sounds that cause them. To an engineer interested in isolating signal they may be considered noise. But to a musician listening back to the sounds just recorded, they can feel more like magic.

No wonder Can turned all the machines on and left the studio, to see what they had to say on their own.

Digital Distortion

There is distortion in digital media too, as Oasis discovered when they pumped up the volume. It can be used as creatively by musicians as any other sound. But I have yet to hear anyone call it magical.

To explain the difference between analog and digital distortion, it's helpful to look at the difference in metering for the two. The VU meter, which reacts slowly and is made to show averages, cannot be used for digital recording. Digital media are linear—they don't perceive sound on a curve, like analog systems (or our bodies). Instead, whatever information is directed at a digital system is the information it receives, 1:1. The meters required for digital recording thus deal in absolutes, not averages, and need a quick response to capture every transient peak because each one is recorded. In the digital studio, there's no tape saturation to compress some peaks into the body of the sound, and no self-erasure to make others disappear.

Digital meters also need to be precise because it's very easy to direct too much information at a digital medium. Anyone with a computer knows that you can't ask a 10 GB hard drive to store 11 GB of information—limits to digital capacity are absolute. What happens to that extra gigabyte of sound, if you try and force a digital audio system to absorb it by turning up to 11, is that it goes missing. Moreover, which gigabyte disappears is not determined by the content of the music. Saturate DAT, as Oasis's producer did, and you don't get harmonic distortions of the musical program, as in analog recording. Instead, you get a distortion of the audio process itself: digital clipping, a hard leveling off of data that results in noise disconnected from musical content.

To guard against this kind of non-programmatic distortion, digital meters are intended to *never enter the red*. Zero on a digital meter is a point beyond which the system does not function. This is nothing (so to speak) like zero on a VU meter, which is meant as an average pivot between a bit too little signal and a bit too much. In fact, the true equivalent of the zero point on a VU meter is approximately −20 on a digital meter. (It is set that low to help ensure that you never reach the "zero full scale" of a digital meter.) [*Figure* 5.5]

When the producer of Oasis's *(What's the Story) Morning Glory?* told his mastering engineer that he wanted the top three LED lights on the meters "lit up ALL the time!" he was pointing to a digital meter, where the last three steps below zero are often colored differently to warn of the approaching absolute cutoff. He

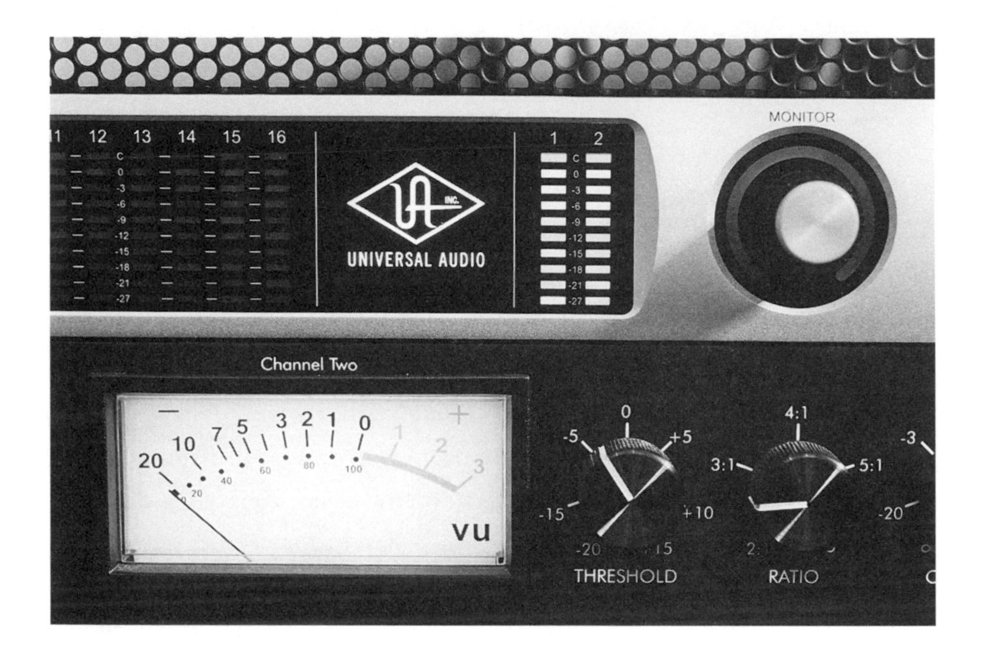

FIGURE 5.5

Above, a digital meter. The topmost level, full scale zero, lights up red to indicate clipping and is not intended to be triggered. Below, an analog VU meter. Zero is a target average reading: the needle is intended to spend at least half the time in the red, maximizing signal-to-noise ratio.

wanted audio information maximized to the limit of the system's functionality.

As the mastered CD attests, the only way to accommodate that request was to bring all sounds up to the level of the loudest peak in the program—not the *average* loudest peak, as in analog, but the *absolute* loudest peak. In other words, make the whole thing one continuous peak. Which means the Oasis CD had zero—true, absolute zero—dynamics. Not to mention a good deal of digital distortion, because at the time the Oasis album was mastered, there were no tools readily available to get that close to the edge without at least sometimes falling over the cliff.

Mute Button

Manipulations of volume in the digital realm extend beyond obnoxiously loud Oasis and Metallica albums. In broadcast media, digital audio tools led to louder and louder television commercials without violating technical limits intended to control their maximum level—until Congress called a stop to the practice by passing the CALM (Commercial Advertisement Loudness Mitigation) Act in 2010.

Metering, again, is at the root of the problem that the CALM Act sought to mitigate. FCC rules developed in the analog era had stipulated that television commercials couldn't exceed the peak volume of the program they interrupted. But analog VU meters, remember, are slow to respond to transient peaks (which don't always survive for transmission via analog media in any case), because they are intended to measure averages rather than absolutes. If the peak on a VU meter for a given program—during the car chase of a procedural, say—measured +2, the audio for a commercial simply couldn't exceed that maximum either.

Enter digital audio recording and the digital meter. The briefest transient peak of the shoot-out in the middle of that chase scene could now be measured precisely and, without breaking the letter of the law, all the audio of the commercial break could be brought up to that brickwall limit. Audio engineers for TV commercials simply followed the same logic as Oasis: keep all the lights on the digital meter lit up all the time. Commercials became one long gunshot.

The results were felt by everyone unable to reach the mute button quickly enough, including Silicon Valley's representative to the House, Anna G. Eshoo. The story goes that Ashoo's elderly father liked to crank the volume on the TV in order to hear the programs better. When the digitally engineered commercials hit, they were "enough to blast me off the couch," as she told the *Wall Street Journal*.[13] One day, with the family over for dinner, she snapped at her brother-in-law to turn the damn thing down during those excruciatingly loud commercials. "You're the congresswoman," he shot back. "Why don't you do something about it?"[14]

The CALM Act passed the Senate unanimously in 2010, a year that will otherwise go down in congressional annals as one of the most divided and contentious of modern political history.

To solve the loudness war in television commercials, the CALM Act mandates a change not only in the rules governing maximum volume but also in metering for broadcast. Television can no longer use the same "full scale db" meter used by digital audio studios. By law it has instead adopted a new metering system which measures LUFS (Loudness Units relative to Full Scale). This loudness unit scale—first developed in Europe for international broadcast associations—applies an algorithm to the linear scale of volume, adjusting for how we *perceive* loudness. Analog beings that we are, that algorithm produces a curve rather than a line, just like tape or any other analog recording medium. In other words, the CALM Act legislates that analog distortion—based on our sense of hearing—must be applied to digital broadcasting.

The new bureaucratic rules for loudness measurement refer to this psychoacoustic curve as "K-weighting." (Hence the FCC's use of a term slightly different from the European LUFS: LKFS, or Loudness, K-weighted, relative to Full Scale.) What the K-weighted scale represents is a rough inverse of our average ability to detect volume relative to pitch, a set of curves plotted initially by a pair of Bell Lab engineers in 1933 as they worked on the intelligibility of speech over the plain old telephone service. The Fletcher-Munson curves, as they are known, have been the basis ever since for understanding how to engineer equal loudness across the hearing spectrum. [*Figure 5.6 & Figure 5.7*]

Harvey Fletcher and W.A. Munson observed that our ears are particularly sensitive to the frequencies around human speech—between 2k and 5k Hz. These show up as a downward bump on the Fletcher-Munson curves and as a slant up on the K-weighting graph. What both indicate is that it takes less volume for our hearing to register loudness in voices, since we are naturally keyed into those frequencies. The new LUFS standard for TV commercials thus tamps those levels down in relation to others.

At the other extreme of our sensitivity, as reflected in both K-weighting and the Fletcher-Munson curves, is our perception of loudness at the lower end of the frequency spectrum. Low tones require a lot more volume to bring them to the same perceived level as higher ones. The LUFS meter thus allows for boosting the lower end of the spectrum—down at the explosions level of a television procedural—since our ears are less sensitive in that region.

But there's an additional crucial observation that Fletcher and Munson made about those low frequencies. The multiple curves on their chart indicate a change in our perception of loudness at different absolute volumes, from the threshold of our hearing at the bottom of their chart to the level of "feeling," aka pain, at the top. And a curious thing happens as we turn up the volume: our perception of loudness begins to level out across the spectrum. The curve moves toward a straight line as volume increases toward the level of pain.

Another way to put this: turn up everything to 11 and you finally begin to hear the bass as loud as everything else. Just as it starts to hurt.

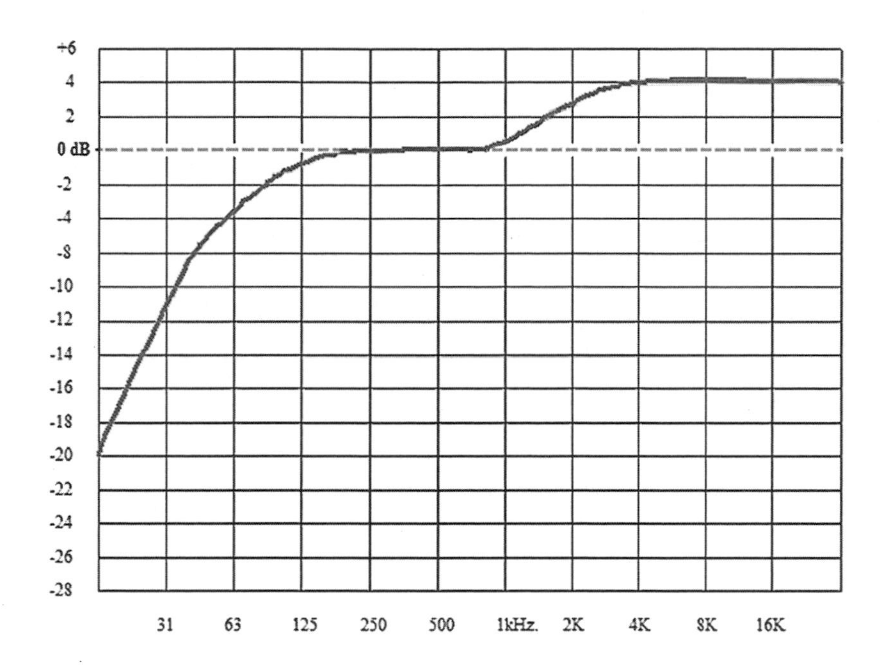

FIGURE 5.6

K-weighting scale across the audible spectrum of sound, from low frequencies (at left) to high (at right).

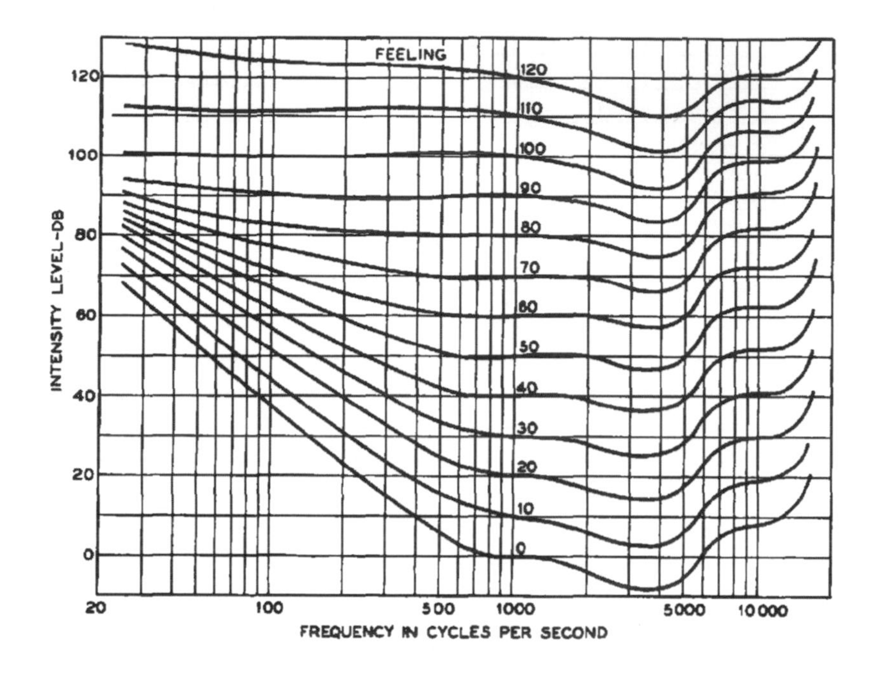

FIGURE 5.7

Fletcher-Munson curves measuring perception of equal loudness, from the *Bell System Technical Journal* (October 1933). At left are lower frequencies; at top are louder volumes. We need more volume to perceive lower tones and less volume to perceive higher ones, especially in the region of speech (the "bump" seen at frequencies between 2000 and 5000 Hz). The curves from bottom to top represent the same test conducted at increasing volume up to the point of "feeling," aka pain.

Bass Drop

While the CALM Act and the switch to LUFS represent an effort to moderate volume in the digital environment for television, in music the loudness wars have opened another front.

The challenge of adequately representing bass on recorded media long bedeviled analog audio engineers. As the Fletcher-Munson curves demonstrate, our ears don't respond to volume equally across the frequency spectrum; we need much more volume to hear lower tones like bass than we do higher ones like our voices. But the isolation of sounds in digital audio recording—and thus their potential manipulation in isolation from one another—makes it easy to "correct" for the Fletcher-Munson curves, and jack the bass up as loud as everything else. Or louder.

Subwoofers are speakers designed to amplify only low-frequency vibrations, that hard-to-reach left edge of the Fletcher-Munson curves. Their first widespread commercial use was in movie theaters for the 1974 disaster film *Earthquake*. Dubbed "sensurround," this new technology consisted of carting subwoofers into theaters and feeding them their own soundtrack of frequencies low and loud enough to make the seats shake. "Please be aware that you will *feel* as well as see and hear realistic effects such as might be experienced in an actual earthquake," the movie posters warned, mirroring Fletcher and Munson's use of "feeling" to describe aural sensation beyond the available sound spectrum.

Then came the carny kicker: "The management assumes no responsibility for the physical or emotional reactions of the individual viewer." [*Figure 5.8*]

Grauman's Chinese Theater in Hollywood hosted the premiere of *Earthquake* and thus of sensurround, even installing a net above the seats to catch plaster falling from the ceiling. Although Grauman's Theater historian Kurt Wahlner doubts any plaster ever actually fell, he attests that, "Having seen *Earthquake* at the Chinese during its first few weeks, I will tell you that it was a great tactic. People didn't know what to expect. And outside of rock concerts, no one had ever heard anything so loud."[15] [*Figure 5.9*]

Unimpressed, *New York* magazine critic Judith Crist dismissed the rumbling as having "all the effectiveness of a drop-a-quarter-in-the-slot motel massage bed."[16] But audiences flocked to it. The "Feelies"—an entertainment dreamed up by Aldous Huxley in 1931 for his dystopic novel *Brave New World*—proved as popular as he'd imagined.

> *The house lights went down; fiery letters stood out solid and as though self-supported in the darkness. THREE WEEKS IN A HELICOPTER. AN ALL-SUPER-SINGING, SYNTHETIC-TALKING, COLOURED, STEREOSCOPIC FEELY. WITH SYNCHRONIZED SCENT-ORGAN ACCOMPANIMENT.*

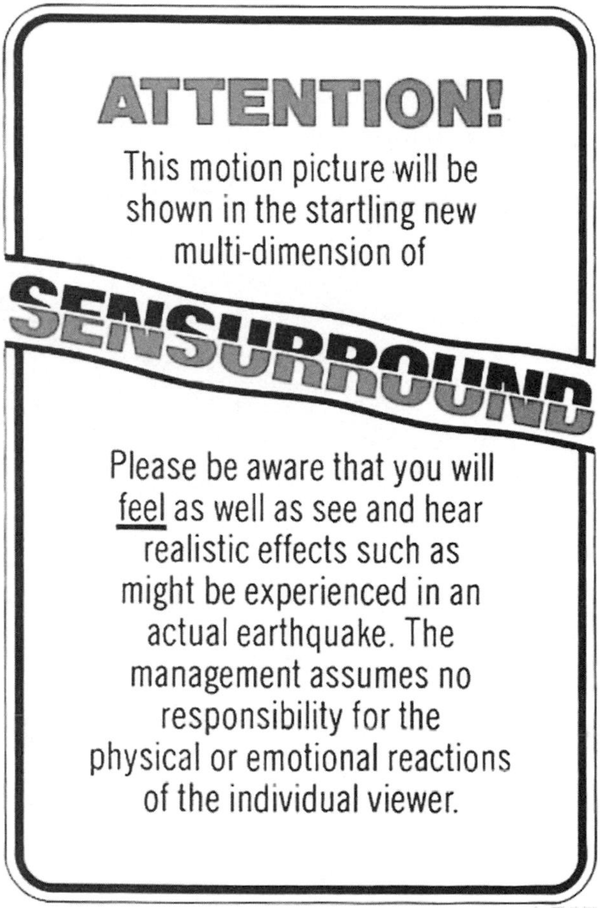

ATTENTION!

This motion picture will be shown in the startling new multi-dimension of

SENSURROUND

Please be aware that you will feel as well as see and hear realistic effects such as might be experienced in an actual earthquake. The management assumes no responsibility for the physical or emotional reactions of the individual viewer.

FIGURE 5.8

Warning to viewers of *Earthquake* (1974).

FIGURE 5.9

Waldon O. Watson, Universal Picture's sound director, with subwoofers installed at the front of Grauman's Chinese Theater for the premiere of *Earthquake* (1974). Watson came out of retirement to help design the sensurround system and received a special technical Academy Award for his efforts.

> *"Take hold of those metal knobs on the arms of your chair," whispered Lenina. "Otherwise you won't get any of the feely effects."*
>
> *The Savage did as he was told.*[17]

Physical and emotional reactions to extremely low frequencies, or "infrasound," have long fascinated military researchers as well as Hollywood special effects departments. In *Extremely Loud: Sound as a Weapon*, Juliette Volcler asserts that much of that military research is, like the warning on those *Earthquake* posters, bunk.[18] Theories about how certain low frequencies could make eyeballs explode, or the existence of a "brown note" that would reduce enemies to "quivering diarrheic messes," have proven to be no more than twisted fantasies. (Sound weapons do exist—witness the Long Range Acoustic Devices deployed by police departments on both Occupy Wall Street and Black Lives Matter protesters—but they operate under different principles than infrasound, chiefly volume.)

There is a medium where twisted fantasies need be no more than twisted fantasies, however: video games. The gaming industry's use of low frequencies combines the cheap Hollywood effect of shaking your seat with the military's dreams of violence, and its tremendous popularity is one of the reasons there are now so many subwoofers in American homes. The Feelies, indeed.

But none of this explains how subwoofers came to be seen as useful for music. In the 1970s, despite the commercial success of

Earthquake, subwoofers made no discernible impression on the music industry—apart from Steely Dan, who had one built to help mix their album *Pretzel Logic* (more twisted fantasies?).[19]

One reason for that is the physical limitations of the LP. Long-playing vinyl records aren't good at reproducing the loud, low frequencies that subwoofers are made to amplify, frequencies that can bounce the needle right out of an LP's microgrooves. Which meant that even *Pretzel Logic*—mastered and pressed as an LP, the dominant format of its day—couldn't benefit commercially from the sound system it had been mixed on.

Twelve-inch vinyl singles, which allow more physical space for their grooves than LPs, can deliver more bass volume, however. Starting in the mid-1970s, 12" singles began to be used for disco, reggae, dub, rap, and bass-driven rock bands like New Order. Originally pressed expressly for dance clubs, these records—unlike the LP or a 7" single—play to effect on systems with subwoofers, and it was in those clubs that the intended "physical or emotional reactions" of such equipment first entered the world of music. At New York's legendary Paradise Garage, the custom-designed "sub-bass" speakers were even named for DJ Larry Levan: the Levan Horn.[20]

Outside a disco, though, in the 1970s you wouldn't find music playing on anything like a Levan Horn. The analog era ended with no one but Steely Dan listening to Steely Dan through a subwoofer.

Pump Up the Volume

As soon as the CD first brought a version of digital audio files into homes and cars in the mid-1980s, the low end started to expand in consumer electronics. The potential for an expanded tonal range was one of the original selling points for the CD—its 1:1 translation of audio was trumpeted as liberation for sound from the distortions of the analog LP. Once the DVD linked digital audio files to video, stereo itself began to cede to 5.1 "surround sound" (not so far from *Earthquake*'s "sensurround," is it?) as a standard for the new consumer ideal of a "home theater."

The 0.1 of that typical home setup is the subwoofer. There's only one because the tones it amplifies are so low that we are unable to locate them; they seem to come from everywhere at once, like the roar of an earthquake. Nevertheless, some choose to install many—like movie theaters did for *Earthquake* and the Paradise Garage did with their Levan Horns. That's not because additional subwoofers add a qualitative difference to the low sounds they generate, but because more power driving more air through more subwoofers generates more sound pressure *on our bodies*. It's about the "feeling" part of the Fletcher-Munson curves, not the hearing ones. This effect may not have been weaponized, as some imagine, but it can be extreme nonetheless.

Guitarist Stephen O'Malley of Sunn O))) is known for low tunings, high volumes, and blocks of physical sound—even his band's name is like an emoticon for soundwaves. His music would seem to be particularly well suited for subwoofers, which is why promoters of a recent U.K. metal festival took it upon themselves to order sixteen of them as part of the band's setup. "Some of the stage security were getting nosebleeds from so much sound pressure!" O'Malley recalled,[21] but probably with more disgust than the festival promoters would have assumed. Their overkill had missed his musical point: "We are actually not trying to create a violent situation, but more of a total immersion of energy."

Nosebleeds at festivals, trance states at dance clubs, intimidation by car audio—multiple subwoofers have their place in the various physical experiences people seek from music. As O'Malley points out, sound pressure *is* energy, and communicating energy can be a part of what music is about.

Yet even for a master of the low-frequency universe like O'Malley, subwoofers don't necessarily add to the qualities of the music he is seeking to share. In fact he doesn't use subwoofers at all for his own listening. "Not in my home," he reports. "A properly set up hi-fi doesn't need a separate sub. If you have well-designed speaker stacks and adequate headroom on the amplifier, it should cover everything. Usually this culture of 'bass boosting' is at the cost of clarity in the rest of the spectrum."

O'Malley is not alone at the detuned end of the musical spectrum. In the early 1990s, guitarist Dylan Carlson and the two-bass lineup of Earth broke low ground in ways that influenced multiple genres of music. Yet it turns out he, too, is no fan of subwoofers.

Indeed, Carlson places subwoofers among "the worst inventions of all time."[22] In response to a series of questions he revealed an anti-subwoofer position as tight as a well-defined bass sound:

Are subwoofers important to you as a musical tool?
"No."
Have you always used them in your live setup?
"No."
Do you use them when you listen to recorded music at home?
"No."
Do you feel subs add an additional musical quality to the low end?
"No."
Are there other artists you look to for creative use of subs?
"No."

If certified low-end experts like O'Malley and Carlson don't use subwoofers to listen to music, why does anyone?

For one, they are already there. The dominance of video and video games in entertainment means that audio in homes is increasingly designed for those experiences, rather than music. As more of us have come to rely on tinny computer speakers, subs have been added to boost the low end of desktop systems. And even though subs can't be incorporated into headphones, Beats by Dr. Dre mimic the effect for portable devices—and they haven't exactly gone broke doing so. In 2014, Apple bought Beats for $3 billion.

From the moment Beats headphones hit the market, audiophiles derided them. "They are absolutely, extraordinarily bad," the editor in chief of *Inner Fidelity* told the *New York Times* in 2011,[23] an opinion you can find repeated ad infinitum by tech reviewers. (One particularly popular YouTube slam called "The Truth About Beats by Dre!" has more than 6 million views.[24]) The problem critics have with Beats is the same described by O'Malley: bass boosting doesn't add to the sound spectrum so much as mask certain frequencies. Beats accomplish their boost in the low end not with clarity but with volume, much like the multiple subwoofers of a system aimed at the body more than the ears.

Jimmy Iovine, the veteran music executive who cofounded Beats with Dre, simply shrugs off the expert naysayers. "The way we hear music is almost the opposite of the way these sound companies hear music," says Iovine.[25] With the purchase of Beats by Apple, that "opposite" way of hearing music may have become the de facto new norm.

The consumer norm for listening to music feeds back into its production. George Clinton, master musician of funk—a genre dependent on big bass—recently returned to the road after a long hiatus. For the first time, he decided to use backing tracks in addition to live instruments, because prerecorded digital audio can add frequencies to the low end that no bass guitar could ever reach. According to an interviewer from *Rolling Stone*, who sat in on rehearsals, "His band hate the idea: It will sound shitty and muffled, they're convinced. Clinton relays their complaints to me without overtly disagreeing, then shrugs. 'That's the sound of today,' he says. 'I don't care how much you think a band is better. The kids are used to that sound, and that's the sound they want to hear. Bass forever! You can hear the bass a block away. So we've got to learn to do it, without copping out completely.'"[26]

Flatline

What the Fletcher-Munson curves tell us is that if you can hear the bass a block away, volume has hit a level where our response to different frequencies flattens out. Everything is louder than everything else.

Loudness is nothing new. Historian Emily Thompson's marvelous website, The Roaring 'Twenties, cross-references noise complaints with Fox Movietone newsreels to construct a cacophonous map of New York City nearly a hundred years ago, when it was evidently just as loud to its contemporaries as it is today.[27] But as Thompson points out in her introduction, sounds alone can't communicate that volume.

> *The aim here is not just to present sonic content, but to evoke the original contexts of those sounds, to help us better understand that context as well as the sounds themselves. The goal is to recover the meaning*

> *of sound, to undertake a historicized mode of listening that tunes our modern ears to the pitch of the past.*[27]

Without context, can we understand loudness? Thompson's argument is about reconstructing the past, but it helps explain why Metallica fans object to *Death Magnetic,* why Stephen O'Malley doesn't use subwoofers at home, and why George Clinton's band groaned when he added digital bass to his live show. Dynamics are a form of context. Without dynamics, we cannot perceive difference in volume, or (as the Fletcher-Munson curves indicate) higher and lower frequencies. When everything is louder than everything else, sounds lose context and thus meaning—even the meaning of loud.

The variable sensitivity of our hearing is itself a form of context for sound. Our particular sensitivity to the range of speech is a distortion of audio information, making noises in that range feel louder even if on an absolute scale they are not. The curves of our perception, you might say, describe our individual contribution to sound, just as the distortions of analog tape describe its contribution to recorded music.

The loudness wars and the bass boosting of digital audio aim to straighten out those curves, pummeling our ears and our bodies with the maximum signal they can absorb. That "sound of today," as George Clinton put it, eliminates the context for noise and reduces our ability to understand it.

6

REAL TIME

ON THE FIRST DAY OF Occupy Wall Street, September 17, 2011, New York City police swept in to Zuccotti Park and . . . confiscated the PA. No amplification without a permit, said the NYPD, pointing to a rather selectively enforced city law. Even bullhorns were forbidden.

Given the level of ambient noise in the financial district of New York, the city authorities must have figured that would be a neat way to drown out the nascent protest. It's difficult to hear someone speaking ten yards away in the neighborhood; addressing a group of hundreds in the open air should be impossible. This was realpolitik loudness war.

But the protestors, in a kind of volume jujitsu, found a way to amplify their voices without electricity. "Mic check!" one would yell. "Mic check!" repeated those within earshot, relaying the cry to a wider circle; and so on, until everyone in the group had heard.

And then the initial speaker would deliver a message a few words at a time, to be repeated again in waves. What became known as the "people's mic" took shape, using the power of multiple voices to rise above the noise of the neighborhood. All of Occupy Wall Street's very democratic General Assemblies—all its debates, all its announcements—would go on to use this system of acoustic amplification. Because there's definitely no law against yelling in New York. [*Figure 6.1*]

Two weeks later, a group in Boston inspired by the action in New York took over Dudley Square, in the shadow of the Federal Reserve Building downtown. Like Zuccotti Park, Dudley Square is a small strip of open land surrounded by some of the busiest traffic in town—a noisy place, especially during the workday. But unlike New York, Boston has no law against using a PA for protest. Organizers set one up against a windowless Transportation Department building at the far end of the square, using its blank wall like a bandshell to project back into the space. The resulting sound was excellent. Yet at Occupy Boston's first General Assembly, the night of September 30, 2011, the group voted to scrap the PA in favor of the people's mic.

As anyone who participated in an Occupy General Assembly knows, communication by people's mic is slow to the point of excruciating. But the constraint of transmitting only a few words at a time—a few words to be repeated—meant that speeches at Occupy were generally concise, meaningful, and urgent to the group. There

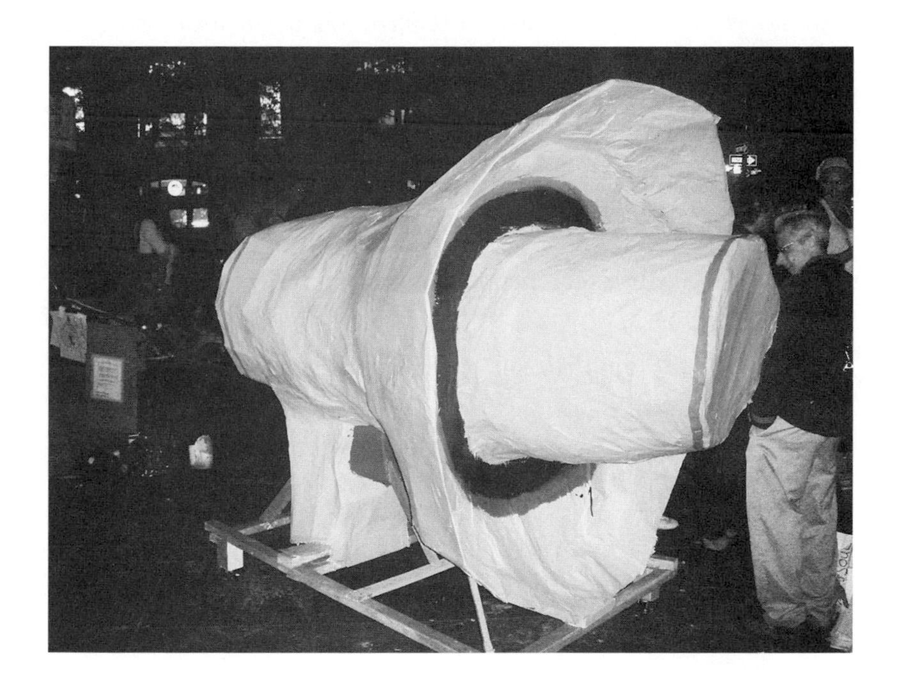

FIGURE 6.1

Papier-mâché bullhorn at Occupy Wall Street.

were very few rambling lectures about off-topic concerns, because no one could bear to repeat them.

The people's mic also proved remarkable in its ability to command attention during meetings; the collective involvement needed to amplify each statement meant that everyone heard them. There was little chatter during General Assemblies, even the dull ones, because all those in attendance were busy making the mic work.

Most significantly, the people's mic decentralized the means of address to the group. Anyone, from any point in the square, had equal access to amplification, rather than just one speaker on a dais. Objections or questions or challenges to a speaker's statement came back at them with the same volume. This decentralization matched the goals and spirit of Occupy precisely.

A historical constraint—imposed on Occupy Wall Street by the NYPD—had led to the development of a means of address reflective of the group putting it to use. The sound of Occupy was a part of its identity.

Rolling

As a young band, Galaxie 500 entered the recording studio like so many before it: completely ignorant of what lay ahead. Analog studios could not easily be approximated at home—there was too much capital and labor involved in building and maintaining one.

Which meant you began learning about them only once the session had already started.

Of course you were likely to have made assumptions about the experience, in ignorance, that proved entirely wrong. Like how to prepare. Who knew you had to play the songs without vocals? Or the rhythm tracks without lead guitar? How could you know in advance what it would be like to hear your bandmates only through headphones, rather than in the room with you? Or how to give a cue without seeing one another?

God forbid the producer insisted on a clicktrack, making the drummer play to a metronome. (Ours didn't. That's how I lived to tell this tale.)

We weren't trying to make a revolution, just a record, but our first day in the recording studio was in this way not unlike the start of Occupy Wall Street: as soon as we arrived, the tools we assumed we would use for the task were taken away. So we had to find others—and fast.

What developed for Galaxie 500 in the studio was what I am sure happened to so many in the same situation. We learned, collectively, how to communicate in that new environment. And the strategies we hit on together became "our" sound, our group identity.

The constraints of the analog studio were, in part, a function of its expense—most musicians, other than the Beatles or Beach Boys, had very limited time in them and had to make the best of it. But regardless of time and money available for a given project, the medium itself demanded that you "commit to tape," as the studio phrase puts it. Decisions in the analog studio were for the most part permanent; little could be undone, short of starting over, because there was no way to move backwards through the process. The sounds you made became your history.

Undo

The invention of analog recording made sound plastic. But it didn't alter time.

In the digital audio studio, history is undoable. Digital audio editing is "nondestructive," which means sounds are saved separately from the operations performed on them. There is no committing to a hard drive. History in a digital audio studio is a tree of commands, which can be pruned back to any point in the past or moved forward again. Sounds are unchanged by decisions, and can be shaped and reshaped ad infinitum.

This malleability reached a new extreme recently with the release of Kanye West's *The Life of Pablo*. After launching the album in the grandest of fashions with a listening party at Madison Square Garden (simulcast to paying customers at an additional seven hundred movie theaters), followed by a global release a few days later on February 14, 2016, Kanye *continued to edit the recording*. He pulled the files from download, making them available through streaming only and thus endlessly revisable.

A month later not only had the tracklist been altered but mixes, lyrics, vocalists, even songwriters and producers had changed. "Life Of Pablo is a living breathing changing creative expression. #contemporaryart," explained Kanye via Twitter as he uploaded yet another version. By "contemporary art," Kanye would seem to mean an art forever in the present—art severed from its own history.

Streaming—the only authorized way to obtain *The Life of Pablo* on its initial release—is a medium for ahistorical listening. On streaming music services like Tidal (Kanye is a part owner) or Apple Music or Spotify, all music is received as "contemporary art" not only because it is all equally available in the present, but because it has been stripped of the information that would tie it to a particular moment of the past.

As Kanye tinkered with his latest, already very public album, Spotify recommended three other "new releases" for me: *Eno: Discreet Music* by Brian Eno, *Box O' Todd (Live)* by Todd Rundgren, and *Donovan—Greatest Hits* by Donovan. Problem is, only one of these recordings is actually new, and it's the one mislisted by artist: this particular recording of *Discreet Music* is not in fact by Brian Eno, who released his (original) version in 1975, but a 2015 interpretation by the Toronto classical ensemble Contact. [*Figure 6.2*]

There is no information on Spotify itself that explains the history of these recordings. They are simply "new," like Kanye's *The Life of Pablo*. Competing streaming platforms are similarly devoid

FIGURE 6.2

Spotify's "New Releases" recommended for the author (March 2016).

of data that would explain the historical constraints that shaped these sounds: the time, the place, the people who recorded them. Not even songwriters are listed—hence the confusion of Contact's version of *Discreet Music* with Eno's own.

The result is a stream of sounds that seem to exist only in the present—the ℗ of recordings, without the © that ties them to history. Which leaves a clear field for an entirely new set of identifying markers.

Pandora's Box

Pandora, the most popular streaming radio service in the United States, launched in 2000 not for consumers at all but as the developer of an information service for other businesses, dubbed (and trademarked) The Music Genome Project®. In keeping with their product's scientific-sounding name, the corporation then known as Savage Beast Technologies emphasized a scholarly approach to their work, although there was more than a touch of Barnum in their claim to "the most sophisticated taxonomy of musical information ever collected."

> *Each song in the Music Genome Project is analyzed using up to 450 distinct musical characteristics by a trained music analyst . . . The typical music analyst working on the Music Genome Project has a four-year*

degree in music theory, composition or performance, has passed through a selective screening process and has completed intensive training in the Music Genome's rigorous and precise methodology.[1]

In practice, this boils down to "trained musicians that come in every day and sit with headphones, entering numbers," as Pandora co-founder Tim Westergren described it elsewhere.[2]

The numbers those musicians are employed to enter form a new set of data for sound recordings, in place of the historical © owned by the record companies. This trademarked ® metadata, in order to be sold for profit, necessarily bypasses information attached to the recordings via liner notes and artwork. Hence the need for a new "taxonomy."

What exactly constitutes that taxonomy—those 450 distinct characteristics that form the "most sophisticated musical information ever"—is private property. Pandora's founders named their product after the public interest Human Genome Project, but they acted closer in spirit to Celera, the privately held company that tried to map the genome for profit.

In fact Pandora wasn't really very much like either of those biotech companies, because there's no such thing as a music genome. Which may be one of the reasons there weren't enough customers for Savage Beast Technologies to make their Music Genome Project® profitable—and why the company decided to change its name in 2004 and create its own consumer product instead:

Pandora Internet Radio. The rest is IPO history. At the close of its first day of publicly traded shares, in June 2011, Pandora had "a market value of $2.8 billion, or about 20 times last year's sales," as *Bloomberg* reported.[3]

Perhaps Pandora found its ideal use for the Music Genome Project® in radio because hidden metadata suits the format. When we listen to radio, we assign a disk jockey the role we might otherwise assume ourselves in choosing what to play. Analog, human DJs use privately held information to select their playlists—dance and rap DJs have been known to tape over record labels to hide the names of their favorite cuts from competitors—so why shouldn't a digital, algorithmic DJ do the same?

There's only one interaction Pandora offers its listeners, after they select a "station" for listening: to click thumbs up or down on tracks particularly enjoyed or not. This ur–social media gesture of up or down judgment—basic to Facebook, Twitter, Instagram, and perhaps most reductively of all the "swipe right/swipe left" on Tinder—is nothing like the engagement of a listener with the information included on an LP or CD. Thumbs up or down is a binary choice: 1 or 0, on or off, yes or no. It is digital logic. And utterly useless for deciphering liner notes, that peculiarly dense and ambiguous form of communication.

Data Points

Terrestrial radio, as the FCC calls the plain old broadcast system, offers the listener no choice at all beyond selecting a station. Internet radio, à la Pandora, adds a binary thumbs up or down for songs—altering a playlist as it unfolds by approving or rejecting its selections. Stream-on-demand services, like Spotify, offer the listener unlimited choice: you can stream any of the music on the service in any sequence at any time. That sequence can follow the structure of an album, like an LP. It can be a playlist of your own creation, like a mixtape. Or it can be someone else's choice, like a DJ.

Given the endless flexibility of streaming services, one would think listeners need vast amounts of information to choose what to hear. It's like walking into the biggest Tower Records imaginable, or the open stacks of Harvard's Widener Library. Everything is there—what now?

But streaming services are anxious about leaving their users in that moment of indecision; endless choice means they might make no choice, and not use the service at all. So instead of supplying copious information for listeners to research their interests—the metadata of printed media—Spotify and the other streaming services have by and large followed suit with Pandora (also iTunes and Napster), and stripped music of all but the bare minimum of tags. In place of supplying information to categorize their massive catalogs, they predict what will interest their users and direct them toward it.

Making effective predictive choices is a science, and some remarkable scientific minds are working on it. Tristan Jehan and Brian Whitman, a pair of MIT PhDs, founded the Echo Nest in 2005 and began applying principles of machine listening and machine learning to the problem of "personalized music discovery experiences."[4] Their data collection puts Pandora's Music Genome Project® to shame: the splash page on the Echo Nest website displays a constantly increasing ticker of "data points" that is well beyond what any number of trained musicians typing for any number of endless workdays could ever enter (it was over a trillion in March 2016). By amassing information about music—they have analyzed more than 37 million songs—and also about the way it is listened to, the Echo Nest truly brings "big data" to bear on what might seem like the rather small problem of making a playlist.

The success of the Echo Nest's methods can be measured in one regard by their client list—Spotify came to rely on them so much that it purchased the company in 2014—or more directly by letting them make suggestions for you. The "Discover" feature on Spotify is a good way of volunteering to be a guinea pig in the Echo Nest lab. A screen grab from my Spotify account's "Top Recommendations for You" presents three albums I do not know, by three artists unrepresented in my personal record collection. And damn if I don't enjoy every one: a female vocal group from 1967, a male soul singer from 1969, and an electric guitar instrumentalist from 2015. [*Figure 6.3*]

How exactly the Echo Nest pinpointed these recommendations

for me is something of a trade secret; the data points collected on them, and on me, are nothing I can consult via my Spotify account or otherwise. Moreover, there's next to no information about any of these albums for me to consult on Spotify. Apart from the cover image, release date, and a two-hundred-word biographical blurb for each of the older artists (there's no bio at all for the contemporary one)—the identical two-hundred-word blurb found on numerous other websites that license such information from the company Rovi—my guinea pig's access to metadata hits a dead end in the maze.

For comparison, consider the metadata included with the original Sweet Inspirations LP: the back cover, not pictured by Spotify. [*Figure 6.4*]

The opening of these liner notes may be empty promotional talk: "When you get an inspiration, and it's a sweet inspiration, what should you call it? A Sweet Inspiration, right? Right!!!" Yet there's much to glean from these notes that situates the recording among a particular group of people, in a particular place and time. We learn that the quartet are experienced backing singers who have already worked together on more than 250 recordings. We learn their names: Cissy Houston (yes, Whitney's mother), Myrna Smith, Sylvia Shemwell, and Estelle Brown. We learn that they formed out of a previous group called the Drinkard Singers, which also included Dionne and Dee Dee Warwick (Cissy's nieces). And we learn that it was Jerry Wexler's idea to name this group and record them as featured artists of their own. (Jerry Wexler was a key

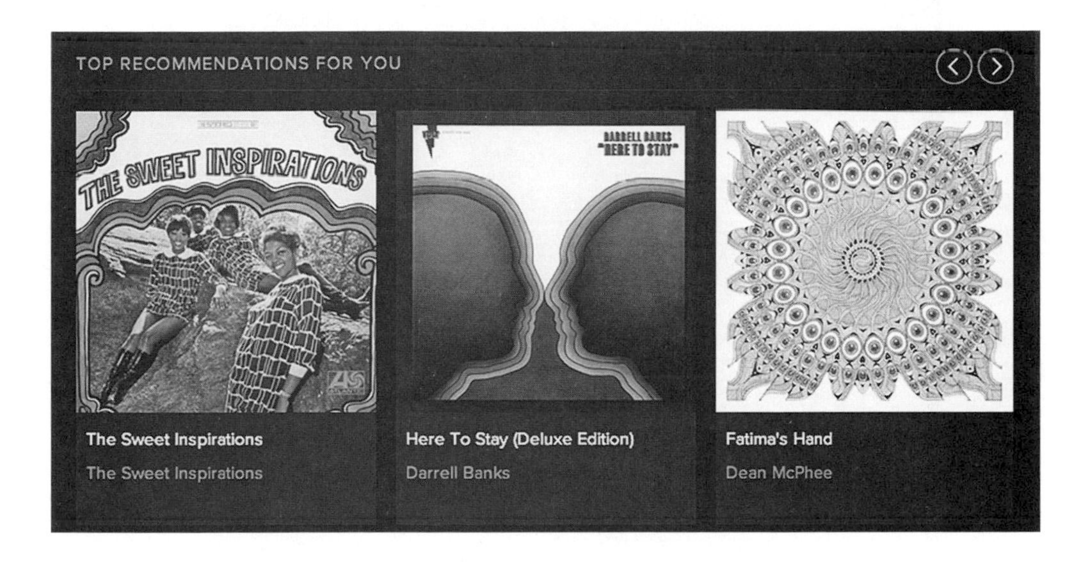

FIGURE 6.3

Spotify's "Top Recommendations" for the author (March 2016).

THE SWEET INSPIRATIONS

When you get an inspiration, and it's a sweet inspiration, what should you call it?

A Sweet Inspiration, right?

Right!!!

And that's what's happening here baby, A Sweet Sweet Inspiration...and how sweet it is!!!

For nearly six years now, musicians, singers, technicians, and the whole backstage scene in the recording studios have been gassed by a tough quartette of soul sisters who have provided the background sounds on more than two hundred and fifty sides—a goodly percentage of them were hits. Producers have been known to postpone important record sessions until "the group" was available to back up their star. And oddly enough, though they have been this important and widely known within the music industry, "the group" never really had a name of its own. But in this groovy business when you're that good, you can't stay "background" forever, there's just too much room out front.

So what happens? Well, the perspicacious Jerry Wexler one fine day said to the delicious Cissy Houston:

"I've got an inspiration."

"Oh, that's sweet!!" replied Cissy after Jerry had spelled out his plan, and the Sweet Inspirations were officially named.

Along with Emily "Cissy" Houston are Myrna Smith, Sylvia Shemwell, and Estelle Brown and like so many of today's great talents these young ladies come loaded with "soul." Each one of them comes from the gospel music field where they have been part of such tremendously popular groups as the Gospelaires and the Drinkard Singers (Dionne and DeeDee Warwick also come from The Drinkards).

And that brings us up to now!

Since that day of inspiration just a short while ago the Sweets have made three single recordings and each one has hit the charts (that's a tough track record any time). This is their initial long playing offering and it's much more than just a package of off-hand out-of-the-can tapes to simply put out a release. This is a gigantic and awe-inspiring creation which well may become the acoustical epitome of soul sounds in our time. Each track is a singular performance by a startlingly "together" group.

Take their unique approach to *Do Right Woman* as a case in point. It takes a lot of heart to come behind Aretha Franklin, but the Sweets decided that they could make it with *Do Right Woman* and with inspired artistry, they turned in an award winning creation! Listen to it!

And for an example of the "now" sound of blues, you couldn't find better than *Blues Stay Away From Me*. But as we indicated earlier, each track in this package is a truly singular performance.

Visually beautiful and artistically immense, The Sweet Inspirations have already won large followings in their theater and night club appearances and will without doubt go on to become one of the top distaff attractions in the business.

Will success spoil the Sweet Inspirations?

No. In addition to their own commitments they still do background work at the rate of about three sessions a week. But this much is true—success for the Sweets has spoiled me, for it's going to take a whole lot of soul to convince me that The Sweet Inspirations aren't the best new group of young ladies in the record business today!

JACK WALKER
WLIB, New York

ATLANTIC RECORDING CORPORATION, 1841 BROADWAY, NEW YORK, NEW YORK 10023

8155
ATLANTIC
MONO

SIDE ONE

1. OH! WHAT A FOOL I'VE BEEN
(By Wallace Pennington, Lindon Dewey Oldham & Darryl Carter; Press, BMI. Time: 2:36)
2. BLUES STAY AWAY FROM ME
(By Alton Delmore, Wayne Raney, Rabon Delmore & Henry Glover; Lois, BMI. Time: 3:22)
3. DON'T LET ME LOSE THIS DREAM
(By Aretha Franklin; 14th Hour-Pronto, BMI. Time: 2:20)
4. KNOCK ON WOOD
(By Eddie Floyd & Steve Cropper; East, BMI. Time: 3:00)
5. DO RIGHT WOMAN—DO RIGHT MAN
(By Dan Penn & Chips Moman; Press, BMI. Time: 3:03)
6. DON'T FIGHT IT
(By Wilson Pickett & Steve Cropper; Cotillion-East, BMI. Time: 2:25)

SIDE TWO

1. SWEET INSPIRATION
(By Wallace Pennington & Lindon Dewey Oldham; Press, BMI. Time: 2:50)
2. LET IT BE ME
(By Mann Curtis, Gilbert Becaud & Pierre Delanoe; Music Corp. of America, ASCAP. Time: 2:41)
3. I'M BLUE
(By Ike Turner; Progressive-Placid, BMI. Time: 2:08)
4. REACH OUT FOR ME
(By Hal David & Burt Bacharach; Jungnickle-Jack-Blue Seas, ASCAP. Time: 2:19)
5. HERE I AM (TAKE ME)
(By David Porter & Isaac Hayes; East, BMI. Time: 3:04)
6. WHY (AM I TREATED SO BAD)
(By Roebuck Staples; Staples, BMI. Time: 2:46)

Horn & string arrangements: Arif Mardin & Ralph Burns

Recording engineers: Chips Moman & Darryl Carter
Cover photo: Stephen Paley
Cover design: Loring Eutemey
Produced by Tom Dowd & Tommy Cogbill
Supervision: Jerry Wexler

This is a high fidelity recording. For best results observe the R.I.A.A. high frequency roll-off characteristic with a 500 cycle crossover.

© 1967 Atlantic Recording Corporation Printed in U.S.A.

FIGURE 6.4

Back cover of the Sweet Inspirations self-titled LP (Atlantic 1967).

figure at Atlantic Records in the 1960s, a man who claimed to have coined the term "rhythm and blues."[5])

That's roughly the same amount of info one might have absorbed from the two-hundred-word bio posted by Spotify, although delivered with a good deal more colorful and outdated slang. However, there's a significant difference: the liner notes provide *names*. Spotify doesn't even tell us which four singers make up the group.

And the names on that back cover don't end with the hepcat liner notes. A track list alongside gives us a flood more: the songwriters Darryl Carter, Lindon Dewey ("Spooner") Oldham, Wallace Pennington, Alton Delmore, Henry Glover, Rabon Delmore, Wayne Raney, Aretha Franklin, Eddie Floyd, Steve Cropper, Chips Moman, Dan Penn, Wilson Pickett, Gilbert Becaud, Mann Curtis, Pierre Delanoe, Ike Turner, Hal David, Burt Bacharach, David Porter, Isaac Hayes, and Roebuk ("Pops") Staples.

Then there are the production credits: arrangers Arif Mardin and Ralph Burns, engineers Chips Moman and Darryl Carter (who had both also snagged a songwriting credit), producers Tom Dowd and Tommy Cogbill, and "supervisor" Jerry Wexler. Not to mention the cover photographer, graphic designer, the music publishers for each songwriting team, the royalty collection agencies for each of those publishers, and finally, with the biggest credit of all and that crucial © to everything above: the Atlantic Recording Corporation.

It took a community of people to make this album, who gathered at a moment in time and shaped a sound together. And none of them made the transition to Spotify, except the random band name explained in that goofy opening to the liner notes. "That's what's happening here, baby. A Sweet Sweet Inspiration . . . and how sweet it is!!!!"

Play-by-Play

Digital streaming services treat the data on the back of an LP—the historical constraints of time, place, and people that created the music—as so much noise. The music alone is filtered through as signal, seemingly always in the present.

But there's a slippage in the present for digital media, as well. "Latency" is the engineer's term for the gap between analog signal and digital information, produced by the time it takes a given computer system to translate between them. It is different for each system, and it is increasingly small (Moore's law again). Nevertheless, it is always there. And there is no way to eliminate it entirely.

Latency is perhaps most obvious in broadcast technology. As television and radio switched from analog to digital transmission, sports fans began to notice something odd: the picture on one no longer lined up with the audio on the other. Here in Boston it was common practice to watch the Red Sox on television with the sound off, listening instead to the play-by-play on radio for the sake of its more voluble and entertaining announcers. Throughout the country there are similar motivations for substituting sports commentary from radio for the television soundtrack—if the television is a national broadcast but the radio commentary remains local, for example.

But the sports fan's solution from analog days doesn't work with digital transmission, and stopped working for everyone in the United States on June 12, 2009, when analog television broadcast was shut down by an act of Congress. Suddenly, radio play-by-play began to describe events before they happened on screen.

One might assume the problem is deliberately caused by television networks, delaying live broadcast for the sake of censorship—to bleep out profanity, or unforseen actions—or perhaps for commercial reasons. But the fact is that some delay in digital broadcast is inevitable; it is an artifact of digital transmission itself. Encoding images to transmit as data and decoding them at the receiving end to display as images again takes time—time that lags behind the real time of events.

Digital radio is equally subject to latency. In the U.K., digital radio users have discovered they can no longer set their clock to the chimes of the hour on the BBC.[6] In Australia, they have learned that they can't bring a digital radio to explain unfolding events at the cricket match.[7] And here in Boston, a home run at Fenway Park is no longer greeted with a simultaneous roar from every open window. Some receive that signal at one moment via cable; some at another via satellite; some at another via an internet connection; and some, quicker than all the others, on a transistor radio. [*Figure* 6.5]

FIGURE 6.5

The author's preferred means for listening to the ball game.

It's not just the different possible paths for digital information that create this lack of simultaneity: different receiving devices have different latency. In my experience, even two iPhones running identical software can fail to synch a streaming digital radio signal arriving over the internet. There are too many variables involved in their individual reception and conversion of the data for there not to be some time difference between them.

Behind the Beat

Latency is no less an issue in the digital recording studio.

Remember how sticky our ears are? In digital audio recording, latency is measured in milliseconds—but so is our sensitivity to sounds in time. Between 30ms and 20ms, we hear latency as echo. Below that is considered "acceptable" by some, but others may find themselves even more sensitive. As the British recording magazine *Sound on Sound* puts it,

> *Musicians will be most comfortable with a figure of 10ms or less . . . Those with golden ears claim to be able to hear shifts of 1ms where "grooves" are concerned.*[8]

"Golden ears" are a figurative commodity in the recording world, although those who convince others they have them can often command gold, quite literally. Regardless, our ears—golden or tin—are a part of our bodies. And even if we aren't one of those who can identify a millisecond shift in the placement of a snare hit within a backbeat, we may well still *feel* it.

Feel (or groove) is precisely what goes awry in digital recording due to latency. Imagine a vertical stack of sounds, each shifted one way or the other by a number of microseconds. At a certain point, the center would become difficult to ascertain. The timescale itself would blur.

In practical terms, latency is a particular problem in the digital studio when overdubbing—the same process that made analog tape such a thick medium for close listening. As a digital recording is played back, it reconverts to an analog signal in order to be heard. Meanwhile, a musician listening to that playback and overdubbing a new track is creating an analog signal that itself requires conversion to digital for recording. These D-to-A and A-to-D conversions take time—latency—which means the *cue for the overdub and the overdub itself don't line up.*

There are of course workarounds. The simplest is to ignore the problem; I hear this on recordings about as often as I see public signage that doesn't use smart quotes. Another crude solution, far more common in professional recording, is to pin everything in the recording to a digital time code. This is the audio equivalent of the "snap to grid" command familiar to many from design programs. A digital recording can be mapped by various divisions of time—beats, frames, samples—and any analog sound arriving into that system can be assigned the nearest place on the grid. The problem is that humans don't play music according to grids, and so every performance needs some adjustment one way or the other. It's like taking that stack of teetering sounds and knocking them against a straightedge to line them up, whether or not they were played that way.

Another workaround, again quite common, is not to use any analog sounds at all. Cue the rise of electronic dance music (EDM), which is intended to move the body but has been played by no actual bodies. This is music made entirely "in the box," as current studio parlance puts it. If all the sounds on the recording originate in the computer, there is no A-to-D conversion and hence no latency.

The final workaround is to adjust the timing of all analog sounds after they are converted to digital, by estimating the particular latency of the system that is being used to record them. This is a standard practice for recording that mixes analog and digital signals, and when executed correctly can result in an apparently seamless reproduction of real time. But "real time" in a digital recording has necessarily been manipulated. It is not the same as real time in the analog sense, which happens with or without our participation.

Digital Time

The digital display clock was invented for a fiction, commissioned by Stanley Kubrick from the Hamilton Watch Company as a prop for his 1968 movie *2001: A Space Odyssey*. Hamilton's designers came up with a clock that featured both a count up and a count down, as if all future time would be keyed to rocket launches. [*Figure* 6.6]

Inspired by their own fantasies, Hamilton went on to develop and market the first digital display watch, the Pulsar, in 1972. New York City private schools being what they were (and still are, even more so I hear), my classmate Steve had one by 1974, when we were in the sixth grade. Everyone was amazed by it. (Steve also had a steady girlfriend before anyone else.)

As a librarian friend pointed out to me many years later, when Harvard's Widener Library trashed their card catalog and she was reaching for a metaphor to explain how the new, digital catalog was different: a digital clock may tell you the precise time *it is*, but an analog clock also shows you all the times *it is not*.

In analog recording, a sound on the tape occupies the place it is and also signifies where it is not. "Music is the space between the notes," is a maxim variously attributed to Debussy, Miles Davis, and any number of other musical geniuses. It is repeated among musicians, and resonates with them, because of this aspect of analog time: we hear time both through signal (the note) and noise (the

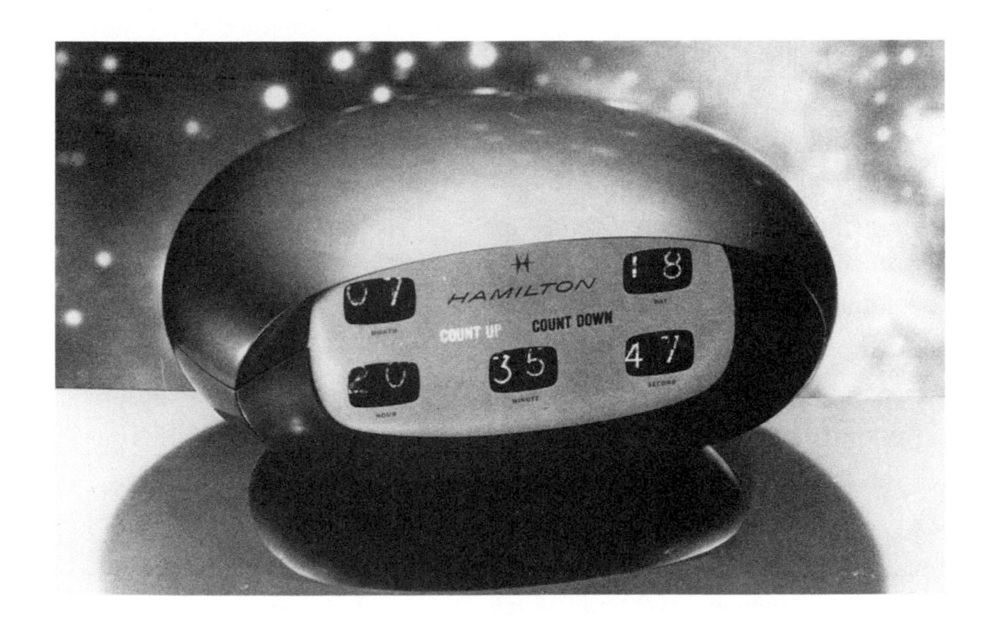

FIGURE 6.6

The Hamilton Watch Company's futuristic clock designed by John M. Bergey and Richard S. Walton as a prop for *2001: A Space Odyssey* (1968). Bergey went on to design the first digital wristwatch, Pulsar, and run Time Computer Inc., the division of Hamilton that manufactured it.

space between the notes). As John Cage taught us, we cannot hear nothing. No space, in analog time, is empty.

Digital time, by contrast, is comprised of *signal only*. The grid used in digital recording is discrete, not continuous—locations on the grid allow for nothing in between. That emptiness between the points of the grid is devoid of signal and of noise. It cannot be touched, not even by our sticky ears.

Digital audio software makes use of the same kind of time displays dreamed up for Kubrick in 1968. In a generic sample window from the audio workstation I use, Digital Performer, the clock is at the top, displayed either in "real time" (minutes and seconds); as a count of bars, like sheet music; of frames, like film; or of samples. Any sound in the recording—represented by the blocks containing waveforms in the window immediately below that display—is pinned to a reading of this clock. But no sound in a digital recording is fixed to any particular place on that clock. Each of those blocks of sound can be individually dragged, nudged, stretched, cut and pasted, or otherwise moved around the timescale at will. [*Figure 6.7*]

In the digital audio workstation, where you are at any moment in the recording is precisely determined by the timescale, but when any given sound occurs is not. All times are equally available to it. Compare this to an analog recording on tape: the tape itself has no absolute time value, but any moment on that length of tape is fixed in relation to all the moments it is not. The sounds at the beginning are necessarily before those in the middle, which precede the sounds at the end.

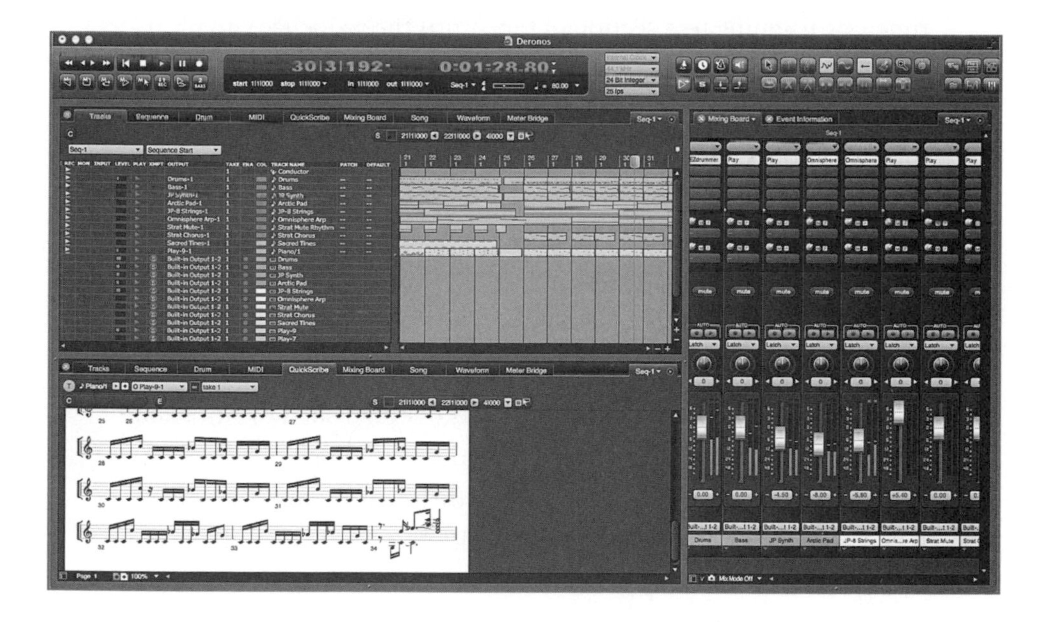

FIGURE 6.7

Screenshot from Digital Performer, a digital audio workstation (DAW) published by Mark of the Unicorn (MOTU).

Relative location is the only way to accurately locate a particular moment on a reel of tape. "Punches" for overdubs in the analog studio—sounds added on the fly, as the tapes are rolling—are timed by listening to a section of tape until the sequence can be anticipated well enough to hit record at the precise moment needed. That moment is distinguished solely by not being the moment before, and not being the moment after.

The Angel of History

Sounds at large are heard the same way a punch is timed in the analog studio, in relative time: the present is distinguished by not being the moment before, and not being the moment after.

Digital audio works with time in a different manner: it is measured against an absolute, discrete scale. No event is fixed relative to any other. And the present is unique to each reception of it, as provisional and changeable as any other point on the timeline.

We do not *feel* digital time so much as we *receive* it. And because we each occupy a unique place with regard to digital information—an individual IP address—we do not receive that time together. *Simultaneity can only be an approximation in digital media.*

When Occupy Wall Street filled the space of Zuccotti Park with signal and noise, its protesters occupied time together. Analog time is shared history—it is lived time, falling into the past at each moment.

Digital time is ahistorical. Its signal arrives without noise—neither the noise of one another nor of the past. It may seem like the future for that reason. But how could we, living in analog time, know what that is? [*Figure 6.8*]

FIGURE 6.8

Paul Klee, *Angelus Novus* (1920).
This drawing was owned by Walter Benjamin and
bequeathed to his friend Gershom Scholem.

This is how one pictures the angel of history. His face is turned toward the past. Where we perceive a chain of events, he sees one single catastrophe which keeps piling wreckage upon wreckage and hurls it in front of his feet. The angel would like to stay, awaken the dead, and make whole what has been smashed. But a storm is blowing from Paradise; it has got caught in his wings with such violence that the angel can no longer close them. This storm irresistibly propels him into the future to which his back is turned, while the pile of debris before him grows skyward. This storm is what we call progress.

—Benjamin, "Theses on the Philosophy of History"

7

LISTENING TO NOISE

THIS BOOK BEGAN BY DRAWING a parallel between the current rush to digital in media, and the postwar redesign of urban life critiqued by Jane Jacobs in *The Death and Life of Great American Cities.* I believe the changes I have experienced in music and examined here indicate that *what we are losing in the demolition of analog media is noise.*

When we listen to noise, we listen to the space around us and to the distance between us. We listen below the surface. We listen each to the limits of our individual perceptions, and we listen together in shared time.

Audio engineers have long sought to isolate and eliminate noise from signal, but only in the digital realm has it become achievable. The signal-to-noise ratio of analog media, no matter how finely engineered, is a relationship that each listener modulates continually. That is: the producer of analog media may determine the universe of noises and signals it contains, but ultimately it's the listener who must sort one from the other. Both signal and noise are always present in analog and both, as we have seen, contain information.

In digital media, signal alone is treated as information, and noise is eliminated. This represents a loss of information—the information communicated by noise. But it is also a change in *who defines signal and who defines noise*. In digital media, the listener receives signal only. The filtering out of noise has already been done by others.

Filtering out noise requires a definition of signal. Whose definition that is—which signal is chosen for isolation, which noise for elimination—is not an engineering problem but a political question. The power to define signal may well be a fundamental struggle in the digital age. So too the power to control signal, once it has been isolated.

This final chapter considers some of the implications of what it means to live—and to try and live as a musician—in a world of signal only.

Information Wants to Be Free, Information Wants to Be Expensive

When Aaron Swartz wrote, "Sharing isn't immoral—it's a moral imperative," he was advocating for the freedom of signal from control, "liberating the information," as he said.[1] The relative inaccessibility of information in analog books and physical libraries didn't trouble him. But once that information was converted into digital signal—purged of noise—its containment, especially for profit, struck him as wrong:

> *The world's entire scientific and cultural heritage, published over centuries in books and journals, is increasingly being digitized and locked up by a handful of private corporations . . . It's time to come into the light and, in the grand tradition of civil disobedience, declare our opposition to this private theft of public culture.*

"Public culture," in Aaron Swartz's use of the term, is analog—information bound to the noise of its physical media. The "private theft" he railed against is digital—the signal of information unbound from noise, freed for virtual communications, and yet, for reasons of power or profit, locked away.

"Information wants to be free," Stewart Brand famously said at the Hackers' Conference in 1984, but also, "Information wants to be expensive . . . So you have these two fighting against each other."[2] In analog media, the signal of information is inextricably bound to noise, making the process of accessing it expensive in terms of time and materials. But in digital media, the signal of information is filtered from noise and in that sense already free. It's not that information's own inclinations are in conflict. The apparent contradiction described by Brand is the result of applying economic terms we accept for analog media to digital signal.

Musicians have been wrestling with this confusion ever since Napster made our digital signals free. In Swartz's terminology, Napster took public culture—music bound to physical media—and freed it for public access.

If you have any sympathy with Swartz's idea of sharing information as a moral imperative, it is hard to take issue with the original peer-to-peer version of Napster that launched in 1999. Listeners shared their sound files with others not for profit but in a free exchange of information. This wasn't piracy or bootlegging, at least not as it was understood then in the world of physical media. There was no manufacturing involved, and more importantly there was no money exchanged. Home taping had established the legality of making a personal copy of recorded music; peer-to-peer exchange didn't seem too far from a giant swap of mixtapes. There wasn't even a blank cassette industry to rail against.

Nevertheless it felt strange to me at first, as I think it did to many musicians, to have one's own music floating around free. Is music free? That simple question provoked by Napster still seems unanswered, all these years of increasing integrated circuit capacity later.

Framing the question in the vocabulary of signal and noise, perhaps we can propose a simple answer: yes, the digital signal of recorded music is or can easily be free, as Napster demonstrated. But that's not *all* the information of music. Napster didn't—indeed it couldn't—convey noise.

Noise doesn't want to be free. Maybe it even wants to be expensive.

Disruption

As the examples in this book illustrate, noise conveys information we use for a full perception of sound, including its location, proximity, and context. A sound file arriving on the desktop via Napster came from anywhere and nowhere, as isolated in space as the woman who fell from her bicycle because headphones removed her from the noise of the street. Similarly, peer-to-peer files arrived from anyone and no one. Although each had been shared by a particular person on the network, there was no way to gauge that user's proximity in physical or social space—they could be in the dorm room next door, or at a computer in an entirely different environment, culture, and time zone.

Napster eliminated these variables for the signals it shared, treating that information as so much noise. And by doing so, it eliminated the expense of shopping for music. Without the artwork, disc, and a shop located in your vicinity, music turned out to be free.

But to benefit from that savings, you had to accept Napster's definition of music as the signal conveyed by a digital sound file.

Do you?

Assigning our individual determination of signal and noise to others is what the digital business community likes to call "disruption."

Napster disrupted the music industry by isolating the digital sound file and equating it with the signal of music, treating everything it could not deliver as noise. A similar strategy allowed Amazon to launch "Earth's Biggest Bookstore" in 1995 without stocking a single title. By isolating the digital catalog listing of a book and eliminating all other context—the quality of its print, the feel of its paper, the store that stocked it, the bookseller who presented it—Amazon was able to deliver books for less money *by providing less information.* The signal of a book was redefined as title, author, and price. The rest was noise, to paraphrase Shakespeare via Alex Ross.[3]

Accessing music in the digital era has similarly shifted from the active search for signals amid noise—browsing in a record store, sitting through other songs until the radio gets to the one you want to hear—toward allowing that search to be made for us by the likes of Pandora and Spotify. This same shift can be seen in any number of disruptive business models. Uber replaces hailing a cab—looking for the signal of a taxi amid the noise of traffic—with an app that sends an isolated car directly to you. Netflix replaces whatever is on at the moment—the noise of television, in shared time—with the isolated signal of video-on-demand. Yelp replaces examining the street around you—that same noisy urban landscape championed by Jane Jacobs—with the isolated signal of a location mapped via GPS.

Disruptive businesses are cocooning us in nothing but signal.

Social Media

Napster's disruption of the music industry was (originally) not for profit. At first it might have been taken as a model for what Aaron Swartz would later call "liberating the information." The digital signal of music, freed from noise, was also free to share.

But it was only a moment until that isolated signal was itself commodified. Successfully sued for copyright infringement by the RIAA, Napster declared bankruptcy and shut down its peer-to-peer system, clearing the way for Apple to launch iTunes as a pay-as-you-go replacement. In place of Napster's copyright anarchy, iTunes established a monopoly over the distribution and pricing of MP3s. Still Apple never tinkered with the mechanics of Napster's isolated sound files: title and artist remained the only information supplied. All that Apple added was a price.

The dizzying success of Apple's iTunes, which quickly came to dominate the sale of music in the United States, and Amazon's bookstore which did the same in publishing, demonstrate the way isolated signal—commodified, even (by some) weaponized—is a locus of digital power and profit. Amazon and iTunes demolished existing retail stores for media as surely as if they were clearing the way for a Robert Moses urban renewal plan. Prior to Napster's launch in 1999, the most recent U.S. Census counted 8,200 record stores. By 2008, just a decade later, only 3,700 remained.[4] The number of independent bookstores has similarly declined by more than half since the launch of Amazon.[5]

What was built in their wake is social media. Facebook launched in 2004; Twitter in 2006; Instagram in 2010.

Social media have no content to offer other than what their users provide. Yet that information, too, is limited to isolated signal as defined by the platform—a neat trick. Take my least favorite of these media, Facebook. Ostensibly, we each define our signals on Facebook by choosing the "friends" whose posts we see. However, Facebook sorts those posts using algorithms to determine which it considers signal—bringing those to our attention—and screening out those that are, in its estimation, noise.

The result is we see nothing but signal on our feeds—that is, nothing but signal as determined by Facebook. Even if its algorithms are accurate, Facebook is screening out precisely the kinds of noise we are skilled at sorting ourselves: posts we don't find especially interesting. But *the act of sorting through that noise is itself a tool for communication*. Just as in audio, sifting social signals from noise helps us gauge distance v. intimacy, and to look beneath the surface of communications.

Without our own active judgment sorting signal from noise, we lose our social bearings on Facebook as surely as the bicyclist wearing headphones lost her sense of location in the street. Routine pleasantries are mistaken for intimate communications, and vice versa. "Friends" in Facebook's terminology are mistaken for friends in our individual estimation, or the other way around. Our emotional communications in social media are so reduced that even the "smiley face"—erstwhile ridiculed emblem of false sentiment—is taken at face value, as it were.

Perhaps most confusingly—and most profitably for Facebook—our own identities are sorted into signal and noise. Information about where we go, what we do, and who we are with is retained by Facebook, stripped of the noise that makes it truly individual, and isolated as signal for sale to those interested in us as data. Unique data points—my 1960s upbringing says we're each one—are outliers and useless in this regard. Useful data points—as all of us are to Facebook—are flattened enough to clump together in marketable chunks. Like the audio loudness wars, our individual signals are amplified and clipped into blocks that can be interpreted at a distance.

We are each an isolated signal to social media, just as its algorithms deliver us nothing but isolated signals.

The Work of Noise

The work accomplished by noise—its communicative power for the maker, its transformative power for the receiver—is missing from both digital media and social media. In digital media, the isolation of signal from noise is a cost-saving measure. In social media, it is a strategy for the monetization of data gathered from users.

It is more than coincidence that this profit-making strategy for the distributors of media has resulted in a loss of income for the producers of its content.

I began writing on these topics because of a surprisingly small royalty check for use of my music by the streaming services Pandora and Spotify. A diminishing ability to make a living in media cuts across disciplines—I might equally have been writing as a journalist, a radio producer, a filmmaker, or (as I am now) an author. The demolition of analog media has been as thorough as it has been quick and has taken with it many of its creators' income streams. New forms of income generated by digital media are heavily weighted toward its platforms rather than its "content providers"—social media being the extreme example, where users are paid zero for producing 100 percent of the content.

However, I believe the examples in this book point to a solution for this dilemma as easy as taking off one's headphones in the street. Just as the disorientation from GPS navigation is solved by paying attention to the analog clues of location all around, the isolation of signal in the digital environment can be corrected by a simple reawakening to the work accomplished by noise. It's the negation of that work that has led to the loss of compensation for its makers.

When we communicate—in private or public, through ephemeral means or by creating the content for media—we generate both signal and noise. Thanks to Moore's Law, in the digital realm signal is more and more easily separated from noise, and more and more quickly put to use. (The net result being that signals in our life are multiplying at an alarming rate.)

But noise remains, always, a part of what we generate in communication. Indeed, thanks to Murphy's Moore's Law, as our signals are increasingly taken up and used for profit by others, noise may be all we have left to call our own.

This is where we take our headphones off. *Noise has value*. It communicates location, proximity, and depth. It tests the limits of our individual perception, and binds us together in shared time. Why shouldn't it want to be expensive?

In fact, noise is more costly than signal—precisely why it is eliminated by disruptive digital-era businesses. Consider the record and bookstores forced out by competition from iTunes and Amazon. Those stores were filled with noise—the noise of unwanted signals. Shelves of product that weren't what you are looking for; customers having conversations that weren't your concern; shop employees offering help yet unable to locate what you did want. On top of that, getting to the shop was a noisy affair costing time, bus fare, fuel, parking . . .

And yet all this noise has a value of its own—the value of shared space and time. The urban spaces we occupy are built on that commonality. The street is a noisy place. And *the street has value*, as Jane Jacobs pointed out to those about to wreck it for superblocks and shopping centers with free parking.

Noisy communications bring that same kind of value to media.

A record purchased in a store is rarely heard just once. It is relatively expensive and will likely be listened to accordingly: on different occasions, with different people, for different reasons. Its sounds will continue to unfold over time, and the signal that record ultimately reveals to its buyer may be quite different from the one it first provided. This deeper, dimensional signal takes shape amid the noise, not despite it. Playing the record contributes to it.

By contrast, a digital music service is designed to deliver isolated signals to isolated individuals, cheaply. If that signal is streaming, it may even be free. And it too will likely be treated accordingly—listened to fleetingly, perhaps not even all the way through, and clicked away into the background as another new file cues up. The user's contribution to the file begins and ends with its download or stream.

There is no denying—Moore's Law—the increased speed, convenience, and low price of digital music delivery. But by Murphy's Moore's Law, it is also an increasingly diminished experience.

It's not uncommon today for consumers to make choices based on qualities other than convenience and price, especially when it comes to fostering community—favoring local stores over chains, for example. If we understand that our media have a similar impact, might we not opt for more noise? Because when we listen to noise, we may well hear each other better.

ACKNOWLEDGMENTS

Liner notes end with credits. It took a lot of people to make this book, though I alone am responsible for its shortcomings. The idea for a study of analog and digital sound was initially suggested by my agent, Alex Jacobs. Astra Taylor read the proposal and recommended it to my editor, Sarah Fan. Carl Bromley guided it through publication at The New Press.

There are thoughts expressed here that I first raised in articles whose editors helped shape and refine them: Ryan Dombal, Mark Richardson, and Lindsay Zoladz at *Pitchfork*; Dan Fox at *frieze*; Jeff Gibson and Don McMahon at *Artforum*.

A Creative Capital | Andy Warhol Foundation Arts Writers Grant supported the research and writing of this book. I also benefitted from the institutional and intellectual support of a fellowship year at the Berkman Klein Center for Internet and Society at Harvard University.

I am very grateful to friends and colleagues who read drafts of the manuscript and offered comments, critiques, encouragement, and invaluable advice for its improvement: Christopher Bavitz, Chris Bohn, Ben Chasny, Larry Crane, Smokey Hormel, Alex Ross, Susanne Sasic, Rob Stenson, Astra Taylor, and Emily Thompson.

Noisy thanks to Naomi Yang, who in addition to all else designed the cover.

NOTES

1. USER'S MANUAL

1. Arthur Salvatore, www.high-endaudio.com. "The Audiophile Website . . . dedicated *exclusively* with the *serious* reproduction of Music."

2. Microtonal composer Ivor Darreg wrote in 1974: "The notes on the staff stand for discrete pitches belonging to some scale or other, and do not truly represent what the human voice or the violin, viola, cello, or bass actually perform from that notation—so writing down a song or violin piece is analog-to-digital conversion, and performing from the notes is digital-to-analog conversion . . . What's so new about digital?" See "Digital, Analog and the Musician," http://www.tonalsoft.com/sonic-arts/darreg/dar12.htm.

3. It has been estimated that in 1986, 99.2 percent of the world's information was stored in analog formats; by 2007, only 6 percent. See Martin Hilbert and Priscila López, "The World's Technological Capacity to

Store, Communicate, and Compute Information," *Science* 332, no. 6025 (April 1, 2011): 60–65.

4. Jane Jacobs, *The Death and Life of Great American Cities* (New York: Random House, 1961).

5. Damon Krukowski, "Making Cents," *Pitchfork*, November 14, 2012, http://pitchfork.com/features/articles/8993-the-cloud/.

2. HEADSPACE

1. Helen Glyde et al., "The effects of hearing impairment and aging on spatial processing," *Ear and Hearing* 34, no. 1, 2013. This article takes issue with whether age itself or hearing impairment with age accounts for the decline in spatial processing after fifty, but in addition to its own findings includes a useful summary of prior studies.

2. "Marcel Proust, Amateur de Theatrophone," Histoire de la Television, January 5, 2002, http://histv2.free.fr/theatrophone/proust1.htm.

3. Steven Watts, *Mr. Playboy: Hugh Hefner and the American Dream* (Hoboken, NJ: John Wiley & Sons, 2009), 125.

4. *The Pet Sounds Sessions* (Capitol, 1997) booklet, 11.

5. See "Dark Side of the Moon on Quad 8 Track," *Quadrophonic Quad*, April 22, 2012, www.quadraphonicquad.com/QQ-PFQ.htm, for details of the quadraphonic *Dark Side of the Moon* release. Flea market denizens should note that the U.S. 8-track tape is highly prized.

6. Alan Parsons, "Four Sides of the Moon," *Studio Sound*, June 1975, www.stereosociety.com/FourSides.shtml.

7. Chris Kyriakakis, "Fundamental and Technological Limitations of Immersive Audio Systems," *Proceedings of the IEEE* 86, no. 5, 1998, quoted in Andria Poiarkoff, "Changes in Spatial Perception Through Headphones," Sonic Arts Research Centre, Belfast, U.K., 2008.

8. W.M. Hartmann, "How We Localize Sound," *Physics Today* 52, no. 11, 1999, quoted in Poiarkoff, "Changes in Spatial Perception."

9. Cheryl Pellerin, "United States Updates Global Positioning System Technology," IIP Digital, U.S. Department of State, February 3, 2006, http://iipdigital.usembassy.gov/st/english/article/2006/02/200602031259 28lcnirellep0.5061609.html.

10. "I Can't Live Without My Radio": Words and music by Rick Rubin and James Todd Smith, 1985.

11. "The New York City Noise Control Code: Not with a Bang, but a Whisper," *Fordham Urban Law Journal* 1, no. 3 (1972): 446–66.

12. CanOpener is an ingenious app that does just this for your iTunes playlist, making it sound like music emanating from speakers instead of earbuds.

3. PROXIMITY EFFECT

1. F.L. Dyer and T.C. Martin, *Edison: His Life and Inventions* (New York: Harper & Brothers, 1910), 38.

2. The app RapidSOS was recently launched to try to solve this locational problem, which the FCC estimates now affects 42 percent of emergency calls. See "A Lifesaving Smartphone App Inspired by a Brush with Tragedy," *New York Times*, September 30, 2015.

3. Apple Keynote iPhone 5 Introduction, September 12, 2012, https://youtu.be/82dwZYw2M00.

4. Bossjock app, "iOS Processing AGC Turned ON for iPhone Built In Mic," *Audioboom*,https://audioboom.com/posts/1550950-ios-processing-agc-turned-on-for-iphone-built-in-mic; and Bossjock app, "iOS Processing AGC Turned OFF for iPhone Built In Mic," *Audioboom*, https://audioboom.com/posts/1550949-ios-processing-agc-turned-off-for-iphone-built-in-mic.

5. According to her colleague and friend Mark Ayres, electronic music composer Delia Derbyshire "used to sit by the phone with a pad making notes as she went along, largely on what she thought you meant through the tone of your voice rather than what you actually said." *Sculptress of Sound: The Lost Works of Delia Derbyshire*, BBC Radio 4 documentary, broadcast March 27, 2010.

6. *Sinatra: An American Original*, CBS Reports documentary, broadcast November 16, 1965.

7. Quoted in Charles L. Granata, *Sessions with Sinatra: Frank Sinatra and the Art of Recording* (Chicago: Chicago Review Press, 2004), 24.

4. SURFACE NOISE

1. John Cage, "Experimental Music," 1957; from *Silence* (Middletown, CT: Wesleyan University Press, 1961), 8.

2. *Temperature's Rising: Galaxie 500, An Oral and Visual History* (Portland, OR: Yeti Books, 2012), 60.

3. Gregory N. Reish, "American Vernacular Music Manuscripts," Center for Popular Music, Middle Tennessee State University, May 11, 2015, http://mtpress.mtsu.edu/popmusic/2015/05/11/american-vernacular-music-manuscripts.

4. John Philip Sousa, "The Menace of Mechanical Music," *Appleton's Magazine* 8, no. 3, 1906.

5. Jonathan Sterne, "Preserving Sound in Modern America," in *Hearing History,* ed. Mark M. Smith (Athens, GA: University of Georgia Press, 2004), 296. See also Sterne, *The Audible Past: Cultural Origins of Sound Reproduction* (Durham, NC: Duke University Press, 2003).

6. Glenn Gould, "The Prospects of Recording" (1966), in *Audio Culture: Readings in Modern Music*, ed. Christoph Cox and Daniel Warner (New York: Continuum, 2004), 117.

7. Curtis Roads, *Microsound* (Cambridge, MA: MIT Press, 2002), 24.

8. Mike Wheeler and Scott Rhoades, "'Don't Talk': Noises and Oddities in Beach Boys Recordings," Cabin, the Web page for Brian Wilson, www.surfermoon.com/essays/noises.html.

5. LOUDNESS WARS

1. Suhas Sreedhar, "The Future of Music: Part One: Tearing Down the Wall of Noise," *IEEE Spectrum*, August 1, 2007, http://spectrum.ieee.org/computing/software/the-future-of-music.

2. "Re-Mix or Remaster Death Magnetic!" GoPetititon, September 10, 2008, https://www.gopetition.com/petitions/re-mix-or-remaster-death-magnetic.html.

3. Mark Yarm, "Metallica Drummer Lars Ulrich Breaks Band's Silence on Death Magnetic Loudness Controversy," September 29, 2008, *Blender*, https://web.archive.org/web/20081003022414/http://blender.com/Blender-Blog-New-Post-09-29-2008/Blender-Blog/blogs/1168/42090.aspx.

4. David Huggins, "Owen Morris: How I Mastered *Morning Glory*," Oasis Recording Information, 2011, www.oasis-recordinginfo.co.uk/?page_id=6.

5. "25 Productions That Made History: Milestones," *Sound on Sound*, November 2010.

6. Ibid.

7. Paul Guy, "Ace Frehley," *Fuzz*, 1997, www.guyguitars.com/eng/interviews/acefrehley.html.

8. Mat Snow, "Reg Presley: 'I Must Learn to Swear More'—a Classic Feature from the Vaults," *The Guardian*, February 5, 2013.

9. Pascal Bussy and Andy Hall, *The Can Book* (Harrow, Middlesex: SAF Publishing Ltd, 1989), 155.

10. Ibid., 98.

11. Ibid., 99–100.

12. Jonathan Sterne, "Preserving Sound in Modern America," in *Hearing History,* ed. Mark M. Smith (Athens, GA: University of Georgia Press, 2004), 315.

13. Daniel Michaels and Elizabeth Williamson, "Well, Hush My Mouth: Congress Is Moving Against LOUD Ads," *Wall Street Journal*, December 1, 2010.

14. Paul Whitefield, "Hugh Hefner Will Like the CALM Act. Other Boomers? Maybe Not," *Los Angeles Times*, December 13, 2012.

15. Email interview with the author, October 2016. See also www.graumanschinese.org.

16. Judith Crist, "Snap, Crackle, Pop," *New York Magazine*, December 2, 1974.

17. Aldous Huxley, *Brave New World* (Garden City, NY: Doubleday, 1932), chap. 11.

18. Juliette Volcler, *Extremely Loud: Sound as a Weapon* (New York: The New Press, 2013).

19. Engineer Ken Kreisel, who went on to co-found the speaker company M&K Sound, called his prototype subwoofer built for Steely Dan, "The Bottom End." He used the name again on a demonstration record for his company's products that later became the secret weapon for a succession of London dance club DJs. See Seymour Nurse's website, www.the bottom end.co.uk.

20. Alan Fierstein and Richard Long, "Paradise Garage Sound System," I Voice, www.ibiza-voice.com/media/news/News/larry_levan/sound.html.

21. Email interview with the author, June 2015.

22. Email and Twitter interview with the author, June 2015.

23. Andrew J. Martin, "Headphones with Swagger (and Lots of Bass)," *New York Times*, November 19, 2011.

24. Marques Brownlee, "The Truth About Beats by Dre!" YouTube, August 30, 2014, https://youtu.be/ZsxQxS0AdBY.

25. Martin, "Headphones with Swagger."

26. Mark Binelli, "George Clinton: Doctor Atomic," *Rolling Stone*, April 27, 2015.

27. Emily Thompson and Scott Mahoy, "The Roaring 'Twenties: an interactive exploration of the historical soundscape of New York City," *Vectors*

Journal 4, no. 1, Fall 2013, http://vectorsdev.usc.edu/NYCsound/777b.html.

28. Ibid., "Author's Statement."

6. REAL TIME

1. Pandora, "About The Music Genome Project®," https://www.pandora.com/about/mgp.

2. John Paul Titlow, "At Pandora, Every Listener Is a Test Subject," *Fast Company*, August 14, 2013, www.fastcompany.com/3015729/in-pandoras-big-data-experiments-youre-just-another-lab-rat.

3. Lee Spears, "Pandora Rises in Biggest Internet IPO Boom Year Since 2000," Bloomberg, June 15, 2011, www.bloomberg.com/news/articles/2011-06-14/pandora-media-raises-234-9-million-in-ipo-after-pricing-stock-above-range.

4. The Echo Nest, "Power of Our Platform," http://the.echonest.com/solutions.

5. Bruce Weber, "Jerry Wexler, a Behind-the-Scenes Force in Black Music, Is Dead at 91," *New York Times*, August 15, 2008.

6. "Digital Time Lag, Time Signals and Delays on TV and Radio," RadioandTelly.co.uk, www.radioandtelly.co.uk/timelag.html.

7. "Digital Radio Delays Play at the Cricket," January 23, 2010,

Radioinfo.com.au, https://www.radioinfo.com.au/news/digital-radio-delays-play-cricket.

8. "Dealing with Computer Audio Latency: Tips & Tricks," *Sound on Sound*, April 1999.

7. LISTENING TO NOISE

1. Aaron Swartz, "Guerilla Open Access Manifesto," July 2008, http://archive.org/stream/GuerillaOpenAccessManifesto/Goamjuly2008_djvu.txt.

2. Stewart Brand, ed., "'Keep Designing': How the Information Is Being Created by the Hacker Ethic," *Whole Earth Review*, May 1985, 44–55, http://tech-insider.org/personal-computers/research/acrobat/8505-a.pdf.

3. Alex Ross's history of twentieth-century music, *The Rest Is Noise* (New York: Farrar, Straus and Giroux, 2007), takes its title from Hamlet's final soliloquy, "The rest is silence." See www.therestisnoise.com/noise.

4. U.S. Census Bureau, Statistical Abstract of the United States: 2000, Domestic Trade and Services, p. 757 (Table 1273. Retail Trade—Establishments, Employees, and Payroll: 1990 and 1997); and U.S. Census Bureau, Statistical Abstract of the United States: 2012, Wholesale and Retail Trade, p. 659 (Table 1048. Retail Trade—Establishments, Employees, and Payroll: 2007 and 2008).

5. "The Amazon Effect" by Steve Wasserman, *The Nation,* May 29, 2012.

NOTES ON THE ILLUSTRATIONS

Figure 2.1 Vintage postcard

Figure 2.2 New York Public Library Digital Collections

Figure 2.11 Photo by Janette Beckman/Getty Images

Figure 3.3 Library of Congress Prints and Photographs Division

Figure 3.4 U.S. Dept. of the Interior, National Park Service, Thomas Edison National Historical Park

Figure 3.5 "Microphone of Caveat April 14, 1877 with mouthpiece added," Library of Congress, Motion Picture, Broadcasting and Recorded Sound Division

Figure 3.9 Photo by Murray Garrett/Getty Images

Figure 4.1 ACT UP New York records, Manuscripts and Archives Division, The New York Public Library Digital Collections

Figure 4.3 Photo courtesy Boston Realty Advisors

Figure 4.5 Courtesy The Pianola Institute, London

Figure 4.6 See *The Story of "Nipper" and the "His Master's Voice" Picture Painted by Francis Barraud,* compiled by Leonard Petts, published by Ernie Bayly for *The Talking Machine Review International*, Bournemouth, England, second edition, 1983

Figure 4.7 Courtesy http://whitefiles.org, archive of the BBC Radiophonic Workshop

Figure 4.8 John Cage Music Manuscript Collection, Music Division, The New York Public Library for the Performing Arts, Astor, Lenox and Tilden Foundations. Courtesy of the John Cage Trust and Edition Peters Group

Figure 5.1 From the online magazine *Recording Hacks*

Figure 5.9 Courtesy Kurt Wahlner and Graumanschinese.org

Figure 6.1 Photo by Ted Sammons

Figure 6.9 Oil transfer and watercolor on paper, 318 cm x 242 cm. The Israel Museum, Jerusalem. Gift of Fania and Gershom Scholem, Jerusalem, John Herring, Marlene and Paul Herring, Jo-Carole and Ronald Lauder, New York. Photo by Elie Posner © The Israel Museum, Jerusalem

ABOUT THE AUTHOR

DAMON KRUKOWSKI was in the indie rock band Galaxie 500 and is currently one half of the folk-rock duo Damon & Naomi. He writes for music and art journals including *Pitchfork*, *Artforum*, *frieze*, and *The Wire*, and is also the author of two books of prose poetry. He is the recipient of an Arts Writers Grant from Creative Capital | Andy Warhol Foundation, and a fellow at the Berkman Klein Center for Internet and Society at Harvard University. Online he can be found at www.DadaDrummer.com.

CELEBRATING 25 YEARS OF INDEPENDENT PUBLISHING

Thank you for reading this book published by The New Press. The New Press is a nonprofit, public interest publisher celebrating its twenty-fifth anniversary in 2017. New Press books and authors play a crucial role in sparking conversations about the key political and social issues of our day.

We hope you enjoyed this book and that you will stay in touch with The New Press. Here are a few ways to stay up to date with our books, events, and the issues we cover:

- Sign up at www.thenewpress.com/subscribe to receive updates on New Press authors and issues and to be notified about local events
- Like us on Facebook: www.facebook.com/newpressbooks
- Follow us on Twitter: www.twitter.com/thenewpress

Please consider buying New Press books for yourself; for friends and family; and to donate to schools, libraries, community centers, prison libraries, and other organizations involved with the issues our authors write about.

The New Press is a 501(c)(3) nonprofit organization. You can also support our work with a tax-deductible gift by visiting www.thenewpress.com/donate.